Reception of Weak Radio Signals from Space

Reception of Weak Radio Signals from Space

Miroslav Kasal
Brno University of Technology, Czech Republic

WILEY

Published by John Wiley & Sons, Inc., Hoboken, New Jersey.
Published simultaneously in Canada.

Library of Congress Cataloging-in-Publication Data applied for

Hardback ISBN: 9781394292738

Cover Design: Wiley
Cover Image: © Sergio Bonilla/Getty Images

Set in 9.5/12.5pt STIXTwoText by Straive, Chennai, India

Contents

List of Symbols

a	radius
a	longer transverse dimension of rectangular waveguides
A	amplitude
A	power transmission
A	counter filling
A_c	carrier amplitude
A_{dB}	gain in dB
A_s	amplitude of the strongest higher harmonic component
AR	axial ratio
b	normalized susceptance
b	attenuation
b	shorter transverse dimension of rectangular waveguides
B	magnetic induction
B	bandwidth
B	count of least significant bits
B_n	noise bandwidth
c	speed of light propagation in free space
C	carrier power
CS	clear sky noise
d	length
d	aperture diameter
D	distance of the Moon from the Earth
D	antenna directivity
D	variance $D = \sigma^2$
D	DAC bit count
D_{eff}	effective DAC bit count
E_b	energy per bit
E_{rms}	effective phase error
E_s	energy per symbol

E_t	decrease of power at the edge of the aperture
E_0	breakdown intensity of electric field in dielectrics
f	focal length
f, F	frequency
F	noise factor (-)
f_b	information (bit) frequency
f_c	cutoff frequency
f_c	clock frequency
f/d	ratio of the focal distance f and the aperture diameter d of a parabolic antenna
F_{dB}	noise figure (dB)
f_{IF}	intermediate frequency
f_o	oscillator frequency
f_{off}	frequency offset
F_{RX}	receiver frequency
$F(s)$	Laplace transform image of a transfer function
F_{TX}	transmitter frequency
$G(\theta, \phi)$	antenna gain
G_r	receiving antenna gain
$G(s)$	Laplace transform image of the transfer function in a direct branch
G_t	transmitting antenna gain
G/T	figure of merit of the receiving system
h	Planck constant 6.63×10^{-34} J s
h	substrate thickness
H_0	DC magnetic field intensity
$H(s)$	Laplace transform image of the transfer function in a feedback branch
h_α	coefficients of piecewise linearized noise characteristics
I_D	drain current
I_{DQ}	quiescent drain current
IM3	third-order intermodulation
$I(s)$	Laplace transform image of a closed loop gain
J_ν	monochromatic brightness
k	Boltzmann constant 1.38×10^{-23} J K^{-1}
k	constant
K_d	conversion gain of the phase detector
K_o	conversion gain of the voltage-controlled oscillator
l	length
L	relative power loss (inverse value of the power transmission A, $A < 1$)

L_{dB}	loss (dB)
m	the first mode number
M	count of symbol bits
M	count of the most significant bits
MN	Moon noise
n	the second mode number
n	frequency multiplication factor
N	number of electrons per unit volume
N	number of bits of the phase accumulator status word FSW
N	noise power
N	frequency divider ratio in the feedback
N_{MI}	noise power at the mixer input
N_{ph}	power spectral density of the phase noise
N_0, N_d, N_ν	noise power spectral density
P	power
P_{1dB}	power at 1dB compression
P_{out}	output power
P_{ph}	phase noise power
P_r	receiver input power
P_t	transmitter power
Q	quality factor
Q_0	quality factor of an unloaded resonator
r, R	radius, distance
R	resistance
R	frequency divider ratio of the reference signal
r_0, R_0	radii of the inner and outer conductor of a coaxial line
R_n	equivalent noise resistance
R_s	high-frequency surface resistance
s	Laplace operator
s	effective reflecting surface of the Moon
s_{ij}	scattering parameters
S	geometric area of an antenna
S_{eff}	effective area of an antenna or of a reflecting surface
S_i	effective aperture of an isotropic radiator
S_Φ	output phase power spectral density
SN	Sun noise
t	microstrip thickness
T	thermodynamic temperature (K)
T_0	ambient temperature
T_A	antenna noise temperature
T_b	brightness temperature

T_{CH}	MMIC channel internal temperature
T_e	equivalent noise temperature
T_G	noise temperature contribution from the Earth
T_{Mo}	Moon brightness temperature
T_R	receiver equivalent noise temperature
T_s	equivalent noise temperature of a receiving system
T_{SKY}	sky brightness temperature
U_0, U_i	root-mean-square value of noise voltage
v	wave propagation velocity
V_{DD}	drain voltage
v_{ph}	phase velocity in a waveguide
w	microstrip width
w_{eff}	effective microstrip width
Y	Y-factor
$y(t)$	relative frequency deviation
Z	impedance
Z_0	characteristic impedance
α	half aperture angle from the focal point of the paraboloid
α	specific attenuation
α_d	attenuation factor of the dielectric in Np m^{-1}
α_{mn}	the *n*th root of the Bessel function of the first kind of the *m*th order
α'_{mn}	the *n*th root of the derivative of the Bessel function of the first kind of the *m*th order
Γ_{opt}	reflection coefficient of the signal source in case of noise matching
Γ_s	reflection coefficient of the signal source
δ	penetration depth
$\Delta\omega(t)$	parasitic frequency modulation
Δ_{TH}	temperature rise of a cooler
ε	effective value of deviations from an ideal paraboloid
ε	permittivity
ε_0	vacuum permittivity
ε_{eff}	effective permittivity
ε_r	relative permittivity
η	efficiency
θ, ϕ	angles in spherical coordinate system
θ_{3dB}	half power beam width
θ_{JC}	MMIC internal thermal resistance
θ_H	cooler thermal resistance
ϑ	Doppler shift

λ	wavelength
λ_0	resonant wavelength of a waveguide
λ_c	cutoff wavelength of a waveguide
λ_g	wavelength in a waveguide
μ	permeability
μ_0	permeability of vacuum
μ_r	relative permeability
ν	frequency
Π	magnitude of Poynting vector (module)
σ	standard deviation
σ	specific conductivity of a dielectric
$\sigma^2(\tau)$	Allan's variance of pairs
τ	interval of determining Allan's pairs
ϕ	harmonic signal phase
Φ	Moon phase
Φ	angle of rotation of the polarization plane by Faraday rotation
$\Phi(s)$	PLL signals in operator form
Ψ	Moon phase delay
Ψ_m	phase reserve of the phase-locked loop
ω	angular frequency
ω_0	oscillator angular frequency, resonant angular frequency
ω_n	natural frequency
ω_p	bandwidth of the phase-locked loop
Ω	spatial angle
Ω_{Mo}	Moon spatial angle in steradians

List of Abbreviations

A/D	analog-to-digital signal conversion
ADPLL	all-digital phase-locked loop
AP	decoding using a priori information
AWGN	additive white Gaussian noise
BER	bit error rate
BPSK	binary-phase shift keying
BWG	beam wave guide
CAT	computer aided transceiver
CFOM	method of Doppler shift compensation (constant frequency on the Moon)
CP	circular polarization
CRC	cyclic redundancy check
CW	continuous wave (telegraphy)
D/A	digital-to-analog signal conversion
DAC	digital-to-analog converter
dBc	power level in dB relative to the carrier level
dBi	antenna gain in dB relative to the isotropic radiator
DBM	double balanced mixer
DC	direct current
DDFS	direct digital frequency synthesizer
DDS	direct digital synthesis
DNL	differential nonlinearity
DPLL	digital phase-locked loop
DSCC	Deep Space Communications Complex
DSN	Deep Space Network
DSP	digital signal processing
DSS	Deep Space Station
ECC	error correction code
EIRP	equivalent isotropic radiated power

EME	communication by lunar surface reflection (Earth-Moon-Earth)
ENR	excess noise ratio
FEC	forward error correction
FF	far field of an antenna
FNF	Fresnel near field of an antenna
FPGA	field programmable gate array
FSK	frequency-shift keying, digital modulation
FSW	frequency setting word of a frequency accumulator
FullDopp	full compensation of Doppler shift by a single corresponding station
GaAsFET	gallium arsenide field-effect transistor
GaN	gallium nitride
GPS	Global Positioning System
GPSDO	GPS-disciplined oscillator
HEF	high-efficiency DSN station
HEM	hybrid electromagnetic wave
HEMT	high-electron-mobility transistor
HGA	high gain antenna
HPBW	half power beam width
HT	Hilbert transform
IMD	intermodulation distortion
INL	integral nonlinearity
IP3	third-order intermodulation intercept point
IQ	quadrature (e.g. mixer)
JPL	Jet Propulsion Laboratory in Pasadena
JT4, JT65	EME digital mode type
LD MOSFET	laterally diffused metal-oxide semiconductor field effect transistor
LDPC	low-density parity check code
LGA	low gain antenna
LHCP	left-handed circular polarization
LNA	low noise amplifier
LO	local oscillator
LP	linear polarization
LPF	low-pass filter
LSB	least significant bit
L1, L2	the first and second Lagrange points, Earth-Sun
MCR	MATLAB compiler runtime
MDS	minimum discernible signal
MGA	medium gain antenna
MMIC	monolithic microwave integrated circuit

MOSFET	metal-oxide semiconductor field-effect transistor
NF	near field of an antenna
NMR	nuclear magnetic resonance
OCXO	oven-controlled crystal oscillator
OP	operational amplifier
PFD	phase-frequency detector
PLL	phase-locked loop
QRA64, Q65	EME digital mode type with LDPC coding
Rb	rubidium
RHCP	right-handed circular polarization
RMS	root-mean-square
RNF	reactive near field of an antenna
s/c	spacecraft
SDR	software-defined radio
SF	solar flux
SFDR	spurious-free dynamic range
SFU	solar flux unit
SiC	silicon carbide
SNR	signal-to-noise ratio
SPLL	software phase-locked loop
SSB	single sideband amplitude modulation
SSPA	solid-state power amplifier
TCXO	temperature-compensated crystal oscillator
TEC	total electron content
TEM	transverse electromagnetic wave
TEmn	transverse electric mode
TMmn	transverse magnetic mode
T/R	transmission/reception
TWT	traveling wave tube
UHF	ultrahigh frequency (decimeter waves)
USO	ultra-stable oscillator
VCO	voltage-controlled oscillator
VHF	very high frequency (meter waves)
WSJT-X	software for EME digital modes
XO	crystal oscillator
XPD	cross-polarization discrimination

Preface

As a 16-year-old boy, I listened to transatlantic tests of Stew Perry W1BB on 160 m in wintertime on very early Sunday mornings. I guessed more than heard his Morse code signals, but found them amazing. As an antenna, I used a 127 m long wire stretched across Toulovcovo náměstí in Litomyšl. And as a receiver, an obsolete E10L served, I retuned to 160. I obtained one of the first OL5AAQ youth licenses at the time, and I could try calling him with 10 W. Once I thought he answered me, but his signals faded into silence. This was my first experience with weak signals.

After studying at the Faculty of Electrical Engineering, Brno University of Technology, I started my internship at the Institute of Scientific Instruments of the Czech Academy of Sciences, where I spent 32 years. In the high-frequency spectroscopy department, I worked on nuclear magnetic resonance (NMR) spectrometers. It is a wonderful specialization where almost all areas of electrical engineering meet. The principle lies in the resonance of atomic nuclei with nonzero spin in a magnetic field. Angular frequency of resonance[1] is given by the relation $\omega = \gamma \cdot B$, where B is the magnetic induction of the field where the core is located, and γ is magnetogyric ratio, which is characteristic for each nucleus. Next to the resolution of the scanned spectra, which is mainly determined by the magnetic field homogeneity, the sensitivity is another basic parameter describing the ability to detect the weakest possible signals. In general, the sensitivity of NMR spectrometers is low compared to other spectroscopic methods. Therefore, all the possible ways for increasing are exploited. Stronger signals can be produced by hydrogen nuclei ^{1}H (protons). Therefore, the so-called CW spectrometers (we started with) recorded the spectra of isotopes ^{1}H, or ^{19}F or ^{31}P mainly. The method was improved by time separation of the irradiation and the detection (so-called "time sharing"). That way, the penetration of the irradiation signal into the receiver was suppressed,

1 Larmor's frequency.

and the sensitivity was not reduced. Nevertheless, chemists hungered for carbon spectra. But the isotope ^{12}C of a large natural occurrence has zero spin, and the isotope ^{13}C with spin ½ is of a very small natural occurrence. In other words, carbon spectra were started to be measured by pulse methods with fast Fourier transform (FFT) [1]. The measurement time was shortened by more than an order of magnitude, the averaging of the repeated response of the spin system was applied, and the sensitivity was increased by an order of magnitude. At the same time, the sensing resonator and the successive low-noise amplifier were instantly optimized. We also considered cooling the resonator at least to nitrogen temperature (77 K). Fundamental advances in both spectra resolution and sensitivity were achieved when using a superconducting magnet ($B = 4.7$ T) developed by another group in our department. I am happy that I could look into these developments as well.

In the 1970s, I also became interested in artificial Earth satellites and space probes. AMSAT (Radio Amateur Satellite Corporation) was established in the USA and Germany. I got involved in their activities, as circumstances allowed me. After the revolution in the 1990s, I was offered the chance to cooperate with AMSAT in Germany intensively, being based at the University of Marburg. I was asked to develop two on-board L-band (1 269 MHz) receivers for the large Phase IIID satellite [3] (see Fig. 1) launched in 2000 from French Guiana into a highly elliptical orbit (HEO). One of those receivers was a part of the main command link that has never been turned off. High sensitivity of both receivers was an imperative.

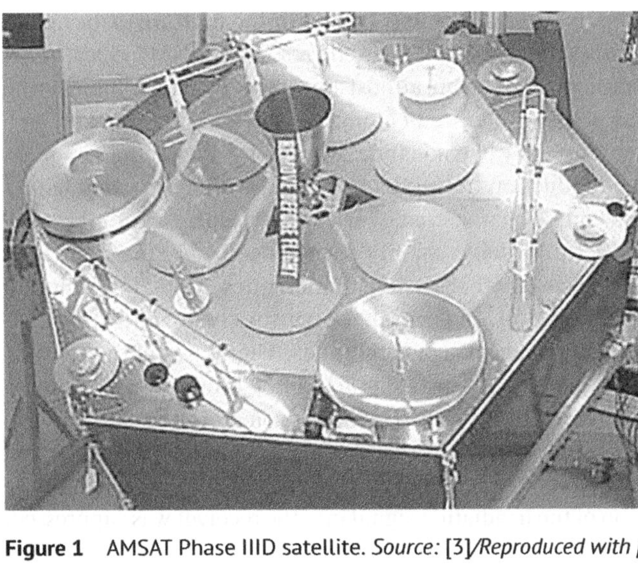

Figure 1 AMSAT Phase IIID satellite. *Source:* [3]/*Reproduced with permission of AMSAT-DL.*

I participated in the satellite integration in Orlando (USA), and the pre-launch campaign in Kourou. Before the launch of the satellite, we built a telemetry and command station at the Department of Radio Electronics, Faculty of Electrical Engineering and Communication, Brno University of Technology (I have since moved). We also managed to create that time unique configuration allowing us to control the station remotely via internet. Operators from the US or Australia were allowed to control the satellite, which was widely used. The satellite was in apogee 60 000 kilometers away. Obviously, weak and very weak signals were still encountered.

In the first decade of this century, we started to cooperate with the US Naval Academy. We jointly designed a transponder for narrow-band keyboard-to-keyboard data communication [2, 4]. A receiver for this project was developed by our lab. The transponder was installed on the ISS for one year (see Fig. 2), and we could perform experiments. Solving the Doppler shift compensation was the main goal. This project was followed by other PSATs, BRICsat, and PSAT-2 projects, continuing to the present day.

In the last decade, I have been attracted by experiments focused on signal reflections from the Moon. Thanks to the group of scientists around Nobel Prize

Figure 2 A narrow-band transponder with our receiver as part of the US Naval Academy SatLab test facility on the ISS. *Source:* [4]/*Reproduced with permission of NASA.*

winner J. H. Taylor,[2] who have been developing software for digital communication by very weak signals, we can communicate by reflection from the Moon (Earth-Moon-Earth, EME) with signals more than 10 dB weaker compared to analog modes. Therefore, we can communicate and experiment with significantly smaller antennas and transmitted powers. This makes the technique more available to a larger number of experimenters. However, the devices still have to meet top parameters to achieve success. Details related to this issue are given in Chapter 12.

So what do "weak signals" mean? Radio signals are called weak if their power level is comparable to the level of natural background noise at a given bandwidth. And we have to emphasize the word "natural," i.e. comparable to the level of thermal noise[3] coming with the signal to the demodulation circuits of the receiver.[4] I can remember the days when the spectrum above 2 GHz was clean and the signal background was really just thermal noise. Unfortunately, such an environment is rarely encountered today. For an experimental work, we have to search for frequencies and locations where those conditions are at least approximately met. Up to frequencies of tens of gigahertz, most radio services operate with signals on the background consisting of other signals or human activity products (the so-called man-made noise). The reasons for this situation are clear and will not be discussed here. People should be aware that the electromagnetic spectrum is a part of the environment and should behave accordingly. National regulators should be more enlightened in this sense, and should take more care of the interests of minorities like in other areas of life in a civilized society.

I have dealt with weak signals for most of my life as a researcher, experimenter, university teacher, and radio amateur. I had the privilege to cooperate with many great colleagues and students. Common stimulating discussions largely inspired me to write this book, which does not aim to be a complete and theoretically exhaustive description of the problem. However, the book tries to briefly summarize diverse practical topics that offer a comprehensive view of the issue of weak signals. And the view is illustrated by certain computational and design examples. Currently, topics related to weak signals are scattered in various professional publications, and concentrating them into one publication seemed useful to me. I will be happy if the book helps the reader to understand the relations, to find their way around this complex issue, and perhaps to like this topic.

---

2 The Nobel Prize winner for the codiscovery of binary pulsars.
3 There are natural noises of other origin as well (shot, 1/f), but these noises can be usually neglected in a narrow-band RF channel.
4 In signal theory, such a channel is denoted AWGN (additive white Gaussian noise).

References

1 Kasal, M., Halámek, J., Švéda, K. et al. (1978). Spectrometer for Fourier spectroscopy of nuclear magnetic resonance (in Czech). *Slaboproudý obzor.* 39: 456–460. ISSN 2336-5773.
2 Bruninga, B., Martinez, P., and Kasal, M. (2002). *PSK-31 Linear/FM Satellite Transponder*. http://aprs.org/psk31uplink.html (accessed 3 January 2023).
3 amsat-dl.org/amsat-phase-3-d-oscar-40/ (accessed 29 July 2024).
4 https://web.archive.org/web/20060525234903/http://www.ew.usna.edu/%7Ebruninga/pec/PC2onISSfullSize.JPG (accessed 29 July 2024).

Acknowledgments

I have almost always been very lucky to have excellent coworkers. A list of their names would be long and I would definitely forget someone. I would therefore like to thank all of them this way; I am truly grateful for their cooperation.

The original Czech edition of this book was reviewed by Professor Miloš Mazánek, PhD, and Jan Šimša, PhD, whom I thank for their valuable comments, discussions, and advice. My great thanks go to Ms. Ivana Jakubová for her meticulous editorial work and comprehensive editing of this book. I would like to thank Professor Dr. Zbyněk Raida for the translation of the book into English and for all his help and support during the publication of the book.

I also thank Mrs. Nandhini Karuppiah, the managing editor of Wiley Publishing, with whom we worked on the English manuscript and who very kindly facilitated our contact with Wiley.

1

Thermal Noise

The power spectral density of thermal noise of a conductor in the radio frequency range, where quantum effects can be neglected, is given by

$$N_0 = k \cdot T \tag{1.1}$$

where $k = 1.38 \times 10^{-23}\ \mathrm{J\,K^{-1}}$ is the Boltzmann constant[1] and T is the thermodynamic temperature of the conductor (in kelvins). If the noise power is measured behind a filter with noise bandwidth B_n, the noise power is

$$N = k \cdot T \cdot B_\mathrm{n} \tag{1.2}$$

In order to understand the relation correctly, let us start from Figure 1.1. Having on the input (power-matched) resistor R of the receiver a noise voltage[2] $U_\mathrm{i} = \sqrt{4kTRB_\mathrm{n}}/2$, the noise power is given by Eq. (1.2). The thermal noise level N (in dBm) as a function of temperature T and bandwidth B_n is shown in Figure 1.2.

Thermal noise is white in the considered part of the spectrum, i.e. the power spectral density is constant. The instantaneous value of the noise varies over time, but corresponding statistical characteristics are constant. The instantaneous values of the noise voltage have a normal (Gaussian) distribution (see Figure 1.3). Such a noise is called AWGN (additive white Gaussian noise). The radio communication channel, where AWGN backgrounds the useful signal, is called AWGN

1 Ludwig Boltzmann used this constant around 1870, long before anyone thought about calculating and measuring electrical thermal noise power. Boltzmann calculated the constant as the energy needed to heat an ideal gas particle by 1 kelvin (the ideal gas particles show elastic collisions only, without an energy transfer). The same constant was used when calculating the entropy. Planck later called it Boltzmann's constant. The value of the constant has been refined over time and is $1.380\,649 \times 10^{-23}\ \mathrm{J\,K^{-1}}$. And the constant has been revealed to be very useful even for "electron gas" in conductors.

2 The result is a mean-square value $U_\mathrm{i} = \sqrt{\overline{u_\mathrm{i}^2}}$, which corresponds to the root mean square (RMS).

Reception of Weak Radio Signals from Space, First Edition. Miroslav Kasal.

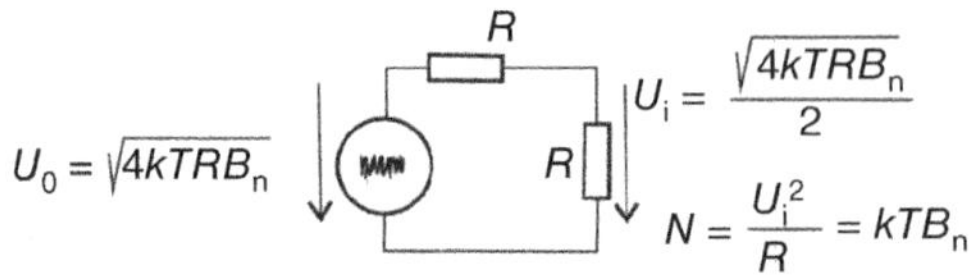

Figure 1.1 Equivalent circuit for calculating the thermal noise power.

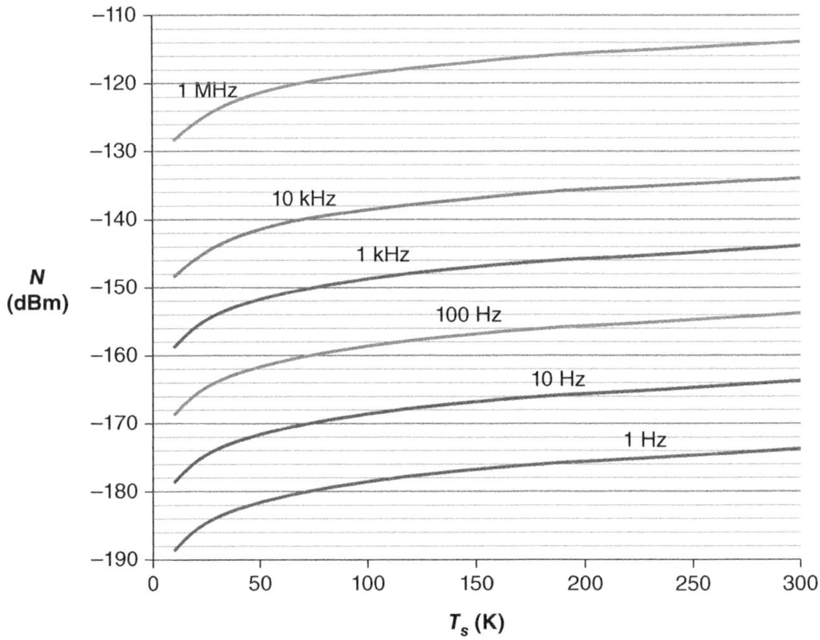

Figure 1.2 The thermal noise level as a function of temperature T and bandwidth.

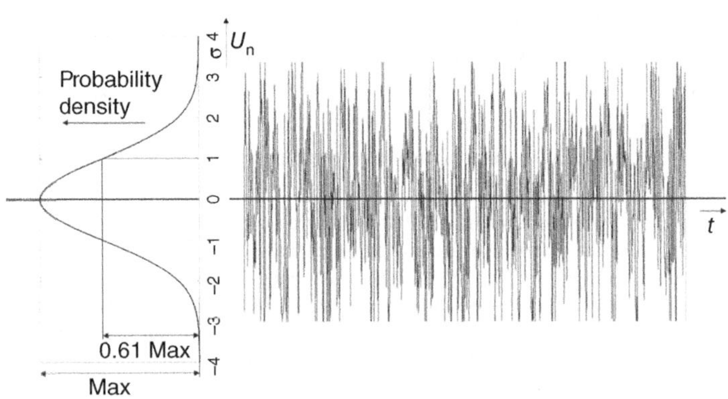

Figure 1.3 Time response of noise voltage U_n and probability density distribution according to the magnitude of instantaneous values. *Source: [1]/with permission of Dalibor Biolek.*

as well. “Additive” means that the signal and the noise are mutually independent, and are adding up in the channel. Separate thermal noise sources (if more sources exist) are mutually independent (uncorrelated) as well, so their powers can be added up. Figure 1.3 shows that the most probable value of the noise voltage is zero, and this is the mean value as well. The standard deviation σ and the variance $D = \sigma^2$ are basic parameters of the normal probability density distribution.

When simulating signal processing processes, the AWGN is characterized as an effective value σ and a noise power D. The peak-to-peak value of the noise voltage can be estimated (with probability of 99%) to be five to six times higher than the effective value.[3]

Reference

1 Biolek, D. *NOISE – Noise Analysis* (in Czech). https://user.unob.cz/biolek/vyukaVUT/prednasky/BMPS/pro_studenty8.pdf (accessed 3 January 2023).

3 And this is the reason why the noise voltage cannot be measured by a standard HF voltmeter. HF voltmeters can usually measure the peak value of the rectified HF voltage, but the scale is given in values that are effective for the harmonic signal, i.e. in the ratio of 0.707. As explained above, this ratio is 0.3 to 0.4 for the noise voltage. This ratio is not only different, but mainly less precisely definable. The noise voltage or the noise power have to be therefore measured by a so-called “true-RMS” voltmeter. Here, the noise voltage is measured by a quadratic voltmeter, or a power-meter based on the thermal effect.

2

Properties of Antennas

2.1 Absolute Black-Body Radiation

An absolute black body is a body that absorbs all the electromagnetic energy (all the wavelengths) with which it is irradiated. The absolute black body is an ideal radiator as well; the body radiates the most energy at a given surface area and temperature. In reality, an absolute black body does not exist, but the radiation of several bodies approaches this limit (e.g. Sun 5 800 K, relic, or CMB, radiation 2.7 K). Planck derived the total power spectral density isotropically radiated from a unitary surface of a black body,

$$N_\nu = \frac{2\pi h\nu^3}{c^2}\frac{1}{e^{h\nu/kT}-1}(\mathrm{W\,Hz^{-1}\,m^{-2}}). \tag{2.1}$$

Here, ν is frequency (Hz), c denotes speed of light (m/s), $k = 1.380\,649 \times 10^{-23}$ J/K is Boltzmann's constant, T is the absolute black-body temperature (K), and $h = 6.63 \times 10^{-34}$ Js is Planck's constant. For radio frequencies with $\nu \ll 2 \times 10^{10} T$ the formula $h\nu \ll kT$ applies and Eq. (2.1) can be replaced by the Rayleigh-Jeans approximation for the lower part of the spectrum:[1]

$$N_\nu \approx \frac{2\pi \nu^2 kT}{c^2} = \frac{2\pi kT}{\lambda^2}(\mathrm{W\,Hz^{-1}\,m^{-2}}). \tag{2.2}$$

Here, λ is the wavelength of radiation. The unit W Hz^{-1} m^{-2} is too high, and therefore the unit[2] 1 jansky (symbol Jy) = 1×10^{-26} W Hz^{-1} m^{-2} was introduced. The Sun's radiation intensity can be expressed using SFUs (solar flux units) 1 SFU = 10 000 Jy. At the frequency of 10 GHz, the typical Sun's radiation intensity is 4×10^6 Jy, while the Milky Way radiation intensity equals to 2×10^3 Jy (see Fig. 2.1).

1 At low frequencies, the quantum nature of the radiation can be neglected. Therefore, h disappears from the relation, and the validity of Eq. (2.1) can be verified that way.
2 The Jy unit was named after Karl Guthe Jansky, who discovered the radiation of the galaxy (1932) and is considered to be the founder of radio astronomy. He was an American physicist (1905–1950) born of Czech immigrants.

Reception of Weak Radio Signals from Space, First Edition. Miroslav Kasal.

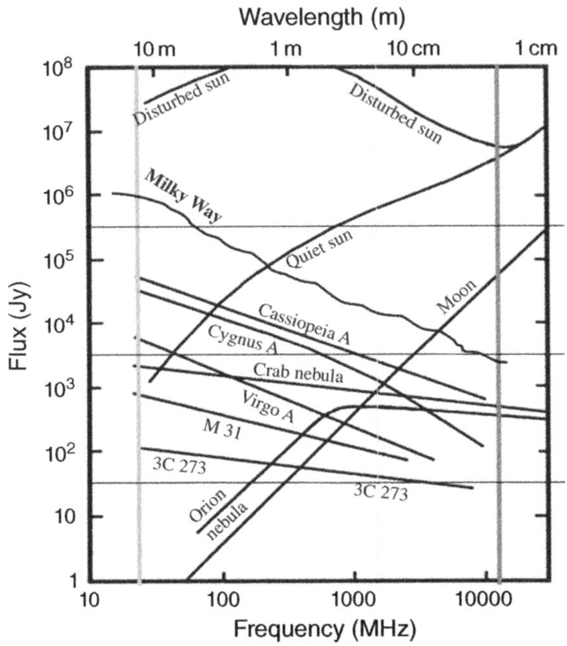

Figure 2.1 Frequency response of radiation level of astronomic objects. *Source: [1]/with permission of National Centre for Radio Astrophysics.*

2.2 Brightness and Noise Temperature of Antennas

When calculating the power spectral density passing through an aperture on a sphere at a distance R (m) from the radiation source that corresponds to the unitary space angle of 1 steradian, a quantity called monochromatic brightness is obtained [1]. Let us consider a spherical black body of radius a. At any point on the surface of a concentric sphere at a distance $R > a$ holds for N'_ν:

$$N_\nu 4\pi a^2 = N'_\nu 4\pi R^2 \quad \text{and so} \quad N'_\nu = \left(\frac{a}{R}\right)^2 N_\nu. \tag{2.3}$$

For an observer on the surface of the outer sphere, the solid angle Ω is defined as $\Omega = \pi a^2 R^{-2}$. For monochromatic brightness, we can therefore write

$$J_\nu = \frac{N'_\nu}{\Omega} = \frac{2\pi \nu^2 kT}{c^2}\left(\frac{a}{R}\right)^2 \frac{R^2}{\pi a^2} = \frac{2\nu^2 kT}{c^2} = \frac{2kT}{\lambda^2} \quad (\text{W Hz}^{-1}\,\text{m}^{-2}\,\text{sr}^{-1}). \tag{2.4}$$

Let us consider an antenna with a narrow beam pattern having a width of $\mathrm{d}\Omega$ (Figure 2.2). This beam intersects the area of the large radiation source, and the effective antenna surface S_{eff} receives a power P_r at the noise channel width B_n. Referring to Eq. (2.4), only half of the power is received since the total radiation

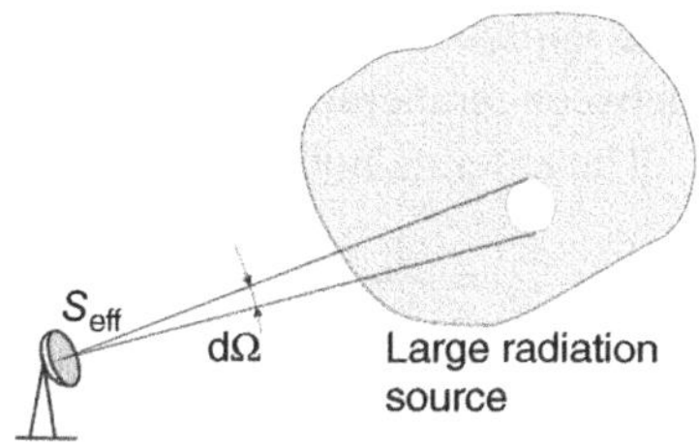

Figure 2.2 Reception of electromagnetic waves from a large source by a directional antenna.

power is divided into two orthogonal polarizations, and the antenna can receive only one polarization. If the antenna is in planes E and H directed at angles (θ, φ) respectively, we can write

$$P_r = \frac{1}{2} J_\nu(\theta, \varphi) \mathrm{d}\Omega\, S_{eff}(\theta, \varphi) B_n = \frac{1}{2} \frac{2kT(\theta, \varphi)}{\lambda^2} \mathrm{d}\Omega\, S_{eff}(\theta, \varphi) B_n. \quad (2.5)$$

Let us consider the antenna theorem. Neglecting ohmic losses of an antenna, we get

$$S_{eff} \cdot \Omega = \lambda^2. \quad (2.6)$$

Here, Ω is the solid angle the antenna radiates the power into. An isotropic radiator is infinitesimally small and radiates into all the directions evenly. Therefore, Ω_i equals to 4π, and the effective surface of the isotropic radiator is

$$S_i = \frac{\lambda^2}{4\pi}. \quad (2.7)$$

The gain of the antenna related to the isotropic radiator is

$$G(\theta, \varphi) = \frac{S_{eff}(\theta, \varphi)}{S_i} = \frac{4\pi}{\lambda^2} S_{eff}(\theta, \varphi). \quad (2.8)$$

The equation for the power received by the antenna, Eq. (2.5), can then be written in the form

$$P_r = \frac{kT(\theta, \varphi) G(\theta, \varphi) B_n \mathrm{d}\Omega}{4\pi}. \quad (2.9)$$

The power received from the entire sphere of a solid angle of 4π steradians

$$P_{rt} = \frac{kB_n}{4\pi} \int_{4\pi} T(\theta, \varphi) G(\theta, \varphi) \mathrm{d}\Omega = kT_A B_n. \quad (2.10)$$

The above cases considered the radiation of an absolutely black body with a thermodynamic temperature T. However, real sources of radiation only approach such a body. In order to reflect this fact, the temperature $T(\theta, \varphi)$ is replaced by the so-called brightness temperature $T_b(\theta, \varphi)$, which is more or less approaching the thermodynamic temperature of the body. The difference between temperatures indicates the emissivity of the real-body surface.

The antenna noise temperature T_A is therefore given by the expression

$$T_A = \frac{1}{4\pi}\int_{4\pi} T_b(\theta,\varphi)G(\theta,\varphi)\mathrm{d}\Omega, \quad \text{or}$$

$$T_A = \frac{1}{4\pi}\int_0^{2\pi}\int_0^{\pi} T_b(\theta,\varphi)G(\theta,\varphi)\sin\theta\,\mathrm{d}\theta\,\mathrm{d}\varphi \tag{2.11}$$

after the coordinate transformation. In the derivation, the antenna is assumed to receive a single polarization only (neglecting cross-polarization). The directional radiation pattern shows a single narrow beam only oriented to the sky. If the side lobes[3] of the antenna radiation pattern are not negligible, the noise temperature T_A has to be evaluated by the following relation

$$T_A = \frac{\int_0^{2\pi}\int_0^{\pi} T_b(\theta,\varphi)\,G(\theta,\varphi)\sin\theta\,\mathrm{d}\theta\,\mathrm{d}\varphi}{\int_0^{2\pi}\int_0^{\pi} G(\theta,\varphi)\sin\theta\,\mathrm{d}\theta\,\mathrm{d}\varphi}. \tag{2.12}$$

Moreover, the radiation of various bodies can represent a desired signal in some cases (in radiometry and radio astronomy, in the calibration of sensitive microwave communication systems) [2]. In communications, this radiation forms the background of transmitted signals. Significant contributions to the noise temperature of the antenna usually come from small areas of space where $T_b(\theta, \varphi)G(\theta, \varphi)$ is the largest, or from large areas of space, where $T_b(\theta, \varphi)G(\theta, \varphi)$ is small [3].

Figure 2.3 shows the frequency response of the brightness temperature T_b of a clear sky for different antenna elevation angles. In clean air, the concentration of water vapor is 7.5 gm^{-3}. An increase of T_b at frequencies around 22 GHz is caused by the increased absorption at the resonance of H_2O molecules.

2.3 Near Field and Far Field

The transmit antenna radiates electromagnetic waves into free space.[4] The generated field varies with the distance from the antenna in the direction of propagation. Two basic zones are distinguished—the near field NF and the far field FF (see Fig. 2.4).

---

3 The radiation pattern side lobes, although considerably weaker compared to the main beam of the antenna, may be directed toward the relatively hot Earth, and their contribution to the antenna noise temperature T_A cannot be neglected.

4 Free space characteristic impedance is 120π ohms (377 Ω). The antenna operates as an impedance transformer between the impedance of the power supply and free space.

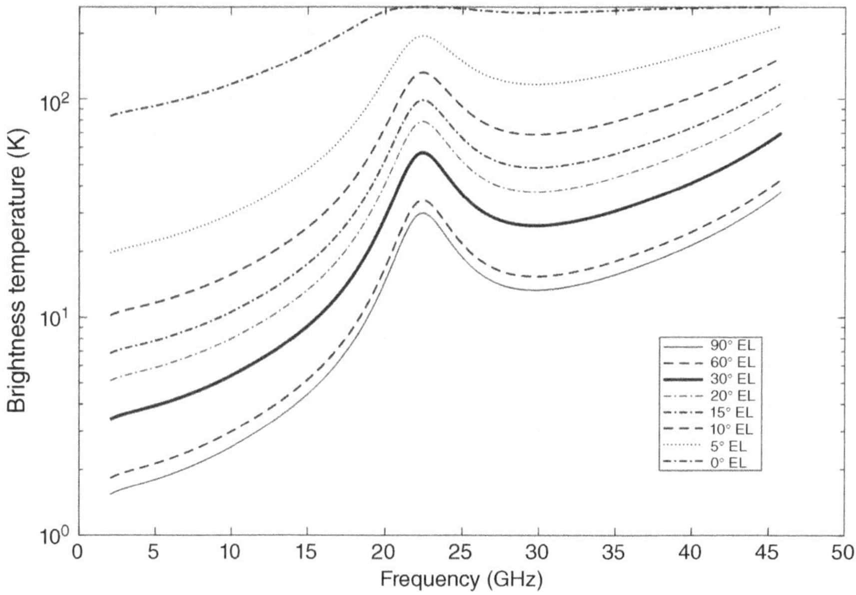

Figure 2.3 Frequency response of the sky brightness temperature for different antenna elevation angles. In a clean atmosphere, the concentration of water vapor is 7.5 gm^{-3}. *Source: 3. HO, C., KANTAK, A., SLOBIN, S. et al. (2007). Link Analysis of a Telecommunication System on Earth, in Geostationary Orbit, and at the Moon: Atmospheric Attenuation and Noise Temperature Effects. IPN Progress Report 42–168.*

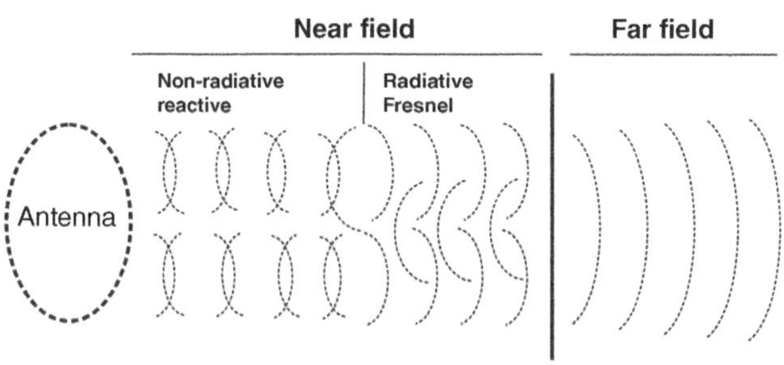

Figure 2.4 Definition of the near field and the far field of the antenna.

Figure 2.4 clearly shows that the near field is divided into a reactive near field RNF (E and H field components are not in phase), and a Fresnel near field FNF. RNF is immediately at the antenna, and FNF is between RNF and FF. The definition of zones is as follows [4]:

$$\text{RNF} < 0.62\frac{\sqrt{d^3}}{\lambda} < \text{FNF} < \frac{2d^2}{\lambda} < \text{FF (m)} \tag{2.13}$$

where d is the largest linear dimension of the antenna. All the following antenna topics are related to the far field (FF). At 10 GHz, an antenna with a parabolic mirror with an aperture diameter of 1.8 m has a far field FF > 216 m from the antenna.[5]

2.4 Directivity and Antenna Gain

When working with weak signals (of an extraterrestrial origin, especially), directional reflector antennas of various configurations are usually used in microwave bands. Antenna directivity $D = D(\theta, \varphi)$ is related to the angular area into which the antenna radiates and also, due to the reciprocity principle, from which it receives radiation. Neglecting other losses (ohmic mainly), the directivity expresses even the gain of the antenna $G(\theta, \varphi) = 10 \log D(\theta, \varphi)$.

For an ideal reflector antenna, the effective surface can be directly related to the surface of the mirror aperture. The effective surface of the aperture S_{eff} is always smaller than the geometric surface of the aperture S since it's impossible for the mirror to be irradiated uniformly (with the same intensity and corresponding phase), a part of the mirror is in the shade of a radiator in a focal point, and a part of the radiated power leaks out of the mirror (spillover)

$$S_{\text{eff}} = \eta \cdot S. \tag{2.14}$$

Here, $\eta < 1$ is the aperture efficiency.

The efficiency is $\eta \approx 0.6$ in case of high-quality and well-adjusted antennas with a parabolic mirror. The spatial radiation pattern of such an antenna ideally shows a narrow main beam of a cigar shape constituting almost all the radiated power. The beam width is measured for 50% power decrease related to the beam center. HPBW (half power beam width) is abbreviated as $\Theta_{3\,\text{dB}}$. If the beam is not rotationally symmetrical, the width is measured in two mutually perpendicular planes $\Theta_{\text{E}3\,\text{dB}}$ and $\Theta_{\text{H}3\,\text{dB}}$ (see Fig. 2.5).

As already mentioned, antenna gain is related to the isotropic radiator and is given by the ratio of effective surfaces. An antenna with a rotationally symmetric parabolic mirror of diameter d has an aperture surface $S = \pi d^2/4$. For S_{eff}, we can write

$$S_{\text{eff}} = \eta \left(\frac{\pi d^2}{4} \right). \tag{2.15}$$

---

5 When measuring radiation patterns of antennas, a considerable space is needed. However, there are methods that allow to derive far-field properties from vector measurements in the near field of the antenna.

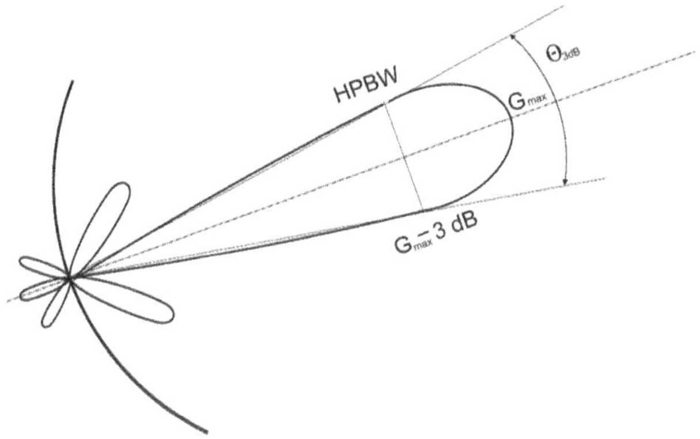

Figure 2.5 Radiation pattern of a parabolic mirror antenna in one plane.

Using Eq. (2.7), the directivity of a parabolic antenna in the center of the main beam is

$$D_{\max} = \frac{S_{\text{eff}}}{S_{\text{i}}} = \eta\left(\frac{\pi d}{\lambda}\right)^2 \text{ and the gain } G_{\max} = 10\log\left[\eta\left(\frac{\pi d}{\lambda}\right)^2\right] \text{ (dBi).} \tag{2.16}$$

The main lobe width of the radiation pattern is proportional to the ratio λ/d. The proportionality constant depends mainly on the efficiency of the aperture and to a lesser extent on other parameters. Practical values of the proportionality constant are in the interval from 64 to 84 and can be found in [5] [6]

$$\Theta_{3\,\text{dB}} = 70\frac{\lambda}{d} \text{ (deg; m, m)} \tag{2.17}$$

and also [6]

$$D_{\max} \simeq \frac{35\,000}{\Theta_{3\,\text{dBE}}\Theta_{3\,\text{dBH}}} \quad (-;\text{ deg, deg}). \tag{2.18}$$

Table 2.1 shows some values calculated from Eqs. (2.16) and (2.17).

The main lobe width affects the noise temperature of the antenna. If pointwise radiators with an angular width α are located in the direction of the maximum gain of a directional antenna without side lobes, their contribution to the noise temperature of the antenna can be expressed as

$$\begin{aligned} \Delta T_{\text{A}} &= T_{\text{b}}\left(\frac{\alpha}{\Theta_{3\,\text{dB}}}\right)^2 && \text{for } \alpha < \Theta_{3\,\text{dB}} \text{ (K; K, deg, deg)} \\ \Delta T_{\text{A}} &= T_{\text{b}} && \text{for } \alpha > \Theta_{3\,\text{dB}} \text{ (K; K)} \end{aligned} \tag{2.19}$$

6 The constant 70 comes from the antenna beam approximation of the Gaussian curve when $\Theta_{3\,\text{dB}}$ = FWHM (full width at half maximum). For an ideally uniform circular aperture, the constant is equal to 58.96.

Table 2.1 Calculated gain values G_{max} and main lobe widths of the radiation pattern $\Theta_{3\,dB}$ for a parabolic antenna with the diameter d and efficiency $\eta = 0.6$.

f/MHz	8 200		10 368		24 048	
d/m	G_{max}/dBi	$\Theta_{3\,dB}$/deg	G_{max}/dBi	$\Theta_{3\,dB}$/deg	G_{max}/dBi	$\Theta_{3\,dB}$/deg
1.2	38.0	2.1	40.1	1.7	47.4	0.7
1.8	41.6	1.4	43.6	1.1	50.9	0.5
2.4	44.1	1.1	46.1	0.8	53.4	0.4
3.0	46.0	0.9	48.0	0.7	55.3	0.3
4.5	49.5	0.6	51.6	0.5	58.9	0.2

while $\Delta T_A = T_b$ corresponds to the situation when the antenna is oriented to a body with a brightness temperature T_b and this body completely fills the cross-section of the radiation pattern.[7]

2.5 Polarization

In general, electromagnetic waves are elliptically polarized. Circular and linear polarization are limiting cases (see Fig. 2.6). The axial ratio AR is the basic parameter. AR = 1 for circular polarization and AR = ∞ for linear polarization. According to an agreed convention, a right-handed circularly polarized (RHCP) wave has an electric component rotating clockwise in the direction of propagation. In the opposite case, a left-handed circularly polarized (LHCP) wave is obtained. The plane of the electric component of the wave has to be defined in case of linear polarization. Horizontal or vertical polarization are most often used for easy orientation. As already mentioned, the antenna can receive only one polarization. Since electromagnetic waves can always be decomposed either into RHCP/LHCP waves or into two orthogonal linear polarizations, a real antenna always receives one half of the power, as shown in Eq. (2.5). A real antenna never achieves a complete suppression of the second polarization. This property is quantified by cross-polarization discrimination, XPD. When transmitting a linearly vertically polarized wave with amplitude A, components of both orthogonal polarizations are received due to the antenna's imperfections (see Fig. 2.7). Cross-polarization

7 Here, we present a simplified theory to clarify basic relations from the practical viewpoint. More detailed quantification of related phenomena can be found in [7] for example.

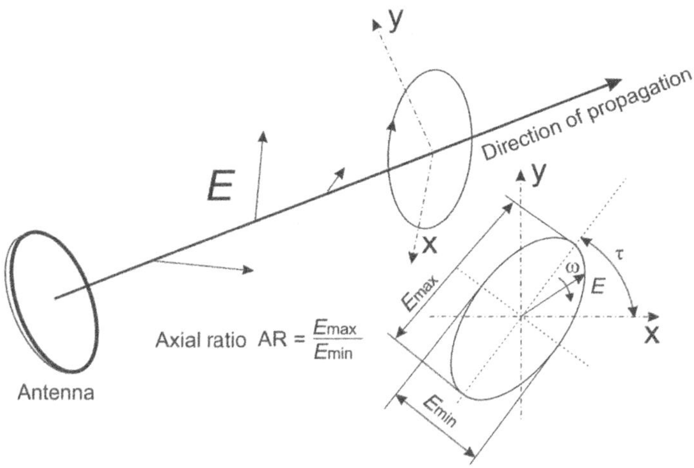

Figure 2.6 Electromagnetic wave polarization. *Source:* [5]/*with permission of John Wiley & Sons.*

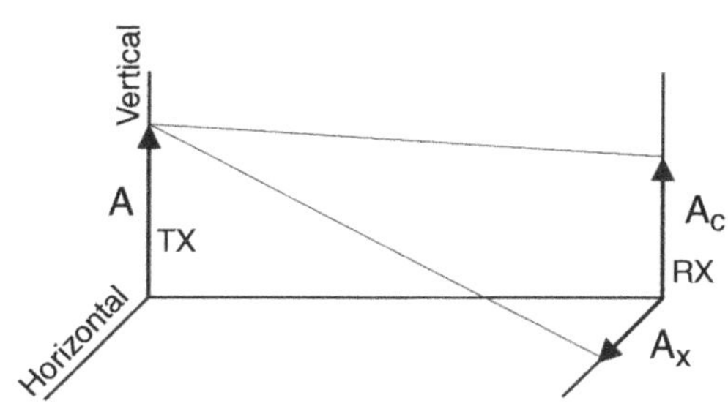

Figure 2.7 Cross-polarization discrimination, XPD.

discrimination, XPD, is then given by

$$\mathrm{XPD} = 20\log\frac{A_{\mathrm{c}}}{A_{\mathrm{x}}}\ \ (\mathrm{dB}). \tag{2.20}$$

For an approximately circularly polarized wave with axial ratio AR, the cross-polarization discrimination is

$$\mathrm{XPD} = 20\log\frac{\mathrm{AR}+1}{\mathrm{AR}-1}\ \ (\mathrm{dB}). \tag{2.21}$$

The axial ratio of a given antenna with elliptical polarization can be experimentally determined by means of cross-polarization discrimination, XPD, as follows [5]:

$$\mathrm{AR} = \frac{10^{\frac{\mathrm{XPD}}{20}}+1}{10^{\frac{\mathrm{XPD}}{20}}-1}. \tag{2.22}$$

Wave components and derived parameters vary with the direction of radiation. Radiation characteristics are therefore given for nominal polarization (co-polar) and for orthogonal polarization (cross-polar). In general, the cross-polarization discrimination, XPD, is highest in the antenna axis and decreases in directions diverging from the direction of the maximum antenna gain.

For linear polarizations, practical values of cross-polarization discrimination, XPD, are about 15 dB. For circular polarization, the values are up to 10 dB higher, i.e. 25 dB. Circular polarization changes the rotation after wave reflection (from RHCP to LHCP and vice versa), which is used to suppress multi-path propagation.

References

1 Koeppen, J. (2013). *Fundamentals of Radio Astronomy*. https://portia.astrophysik.uni-kiel.de/~koeppen/10GHz/basics.html (accessed 7 August 2024).

2 Otoshi, T.Y. (2008). *Noise Temperature Theory and Applications for Deep Space Communications Antenna Systems*. Artech House.

3 Ho, C., Kantak, A., Slobin, S. et al. (2007). Link analysis of a telecommunication system on Earth, in geostationary orbit, and at the Moon: atmospheric attenuation and noise temperature effects. *IPN Progress Report* 42: 168.

4 Editorial Team. (2018). *Everything RF*. https://www.everythingrf.com/community/what-are-near-field-and-far-field-regions-of-an-antenna (accessed 7 August 2024).

5 Maral, G. and Bousquet, M. (1993). *Satellite Communication Systems*. 2nd ed. New York: Wiley.

6 Raida, Z. and Hanus, S. (2007). *High Frequency Technology and Antennas* (in Czech). https://docplayer.cz/69218841-Vysokofrekvencni-technika-a-anteny.html (accessed 7 August 2024).

7 Balanis, C.A. (2005). *Antenna Theory*. 3rd ed. New Jersey: Wiley.

3

Two-Port Network Noise Figure and Equivalent Noise Temperature

For a linear two-port network that is matched at the input and the output, we can define the noise factor as

$$F = \frac{S_1/N_1}{S_2/N_2} \text{ and the noise figure } F_{\text{dB}} = 10\log F. \tag{3.1}$$

In other words, the noise factor is signal-to-noise ratio S/N at the input of the two-port network related to the same ratio at the output. The noise factor is always higher than 1 since a real two-port network can only reduce the signal-to-noise ratio.

Next to the noise figure, the equivalent noise temperature, T_e, is used to quantify noise parameters of radio frequency (RF) systems. Noise power (AWGN) at the input of an amplifier with noise figure F and gain A (see Fig. 3.1), if power matching and noise bandwidth B_n are considered, equals $N_1 = kT_0B_n$. Here, T_0 is the thermodynamic temperature of the signal source with internal resistance R_g. For calculating the noise factor (noise figure), IEEE has determined the temperature $T_0 = 290$ K [1, 2]. Modifying Eq. (3.1), we get

$$F = \frac{S_1}{S_2}\frac{N_2}{N_1} = \frac{S_1}{S_1A}\frac{kT_0B_nFA}{kT_0B_n}. \tag{3.2}$$

In accordance with the model in Figure 3.1, a real two-port network with noise factor $F > 1$ can be substituted by an ideal (noiseless) two-port network with $F = 1$ that is supplemented by a second additive input. To this input, an identical resistor R_g is connected, but the temperature T_e corresponds to the same noise power on the load

$$N_2 = (kT_0B_n + kT_eB_n)A = kT_0B_nFA \tag{3.3}$$

and from here

$$F = \frac{T_0 + T_e}{T_0}, \quad T_e = T_0(F - 1). \tag{3.4}$$

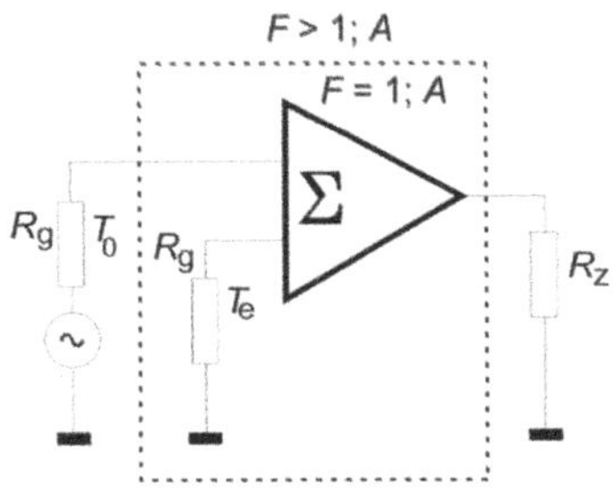

Figure 3.1 The physical meaning of the equivalent noise temperature.

For RF systems with a low level of intrinsic noise, the noise temperature enables sufficiently accurate description of noise properties, and the influence of separate components is clearer. An expression for the resulting noise factor of n cascaded and matched two-port networks, the so-called Friis relation, is:

$$F = F_1 + \frac{F_2 - 1}{A_1} + \frac{F_3 - 1}{A_1 A_2} + \dots + \frac{F_n - 1}{A_1 A_2 \dots A_{n-1}}. \tag{3.5}$$

Using equivalent noise temperatures, we obtain

$$T_e = T_{e1} + \frac{T_{e2}}{A_1} + \frac{T_{e3}}{A_1 A_2} + \dots + \frac{T_{en}}{A_1 A_2 \dots A_{n-1}}. \tag{3.6}$$

According to Eq. (3.6), the resulting noise temperature of a two-port networks cascade cannot be lower than the noise temperature of the first element. Moreover, the contribution of each subsequent element is weighted by the gain of previous elements.

3.1 Noise Figure of Radio Frequency Attenuator

RF attenuator[1] is a two-port network consisting only of resistors with temperature T_F. If L_F are relative signal losses caused by the attenuator that is matched both at the input and the output, then the equivalent noise temperature of the attenuator is

$$T_e = T_F(L_F - 1). \tag{3.7}$$

If $T_F = T_0$, the noise factor of the attenuator is

$$F_F = L_F = \frac{1}{A_F} \text{ and the noise figure of the attenuator } F_{F\,dB} = -10 \log A_F. \tag{3.8}$$

Here, A_F is the power transfer by the attenuator.

1 An RF attenuator usually consists of three resistors in a star (T-cell) or delta (π-cell) configuration. The input and output impedances are usually equal to the characteristic impedance of the line the attenuator is connected to.

3.2 Noise Bandwidth

A communication channel is usually defined by the bandwidth of the transfer characteristic for a decrease of 3 dB (see Fig. 3.2). Since the selectivity curve is not ideally rectangular, the integral noise level in the channel is higher than the noise corresponding to $B_{3\mathrm{dB}}$ (the noise outside the passband is applied as well). Therefore, we introduce the noise bandwidth B_n [2], which corresponds to the ideal selectivity curve with the same additive noise power at the channel output. If $A(f)$ is the channel power transmission, then

$$B_\mathrm{n} = \frac{1}{A_\mathrm{m}} \int_0^\infty A(f)\,\mathrm{d}f. \tag{3.9}$$

3.3 System Noise Temperature

The microwave receiving system (see Fig. 3.3) consists of an antenna with noise temperature T_A, feeder with losses L_FRX and thermodynamic temperature T_F, and the receiver itself with equivalent noise temperature T_R.

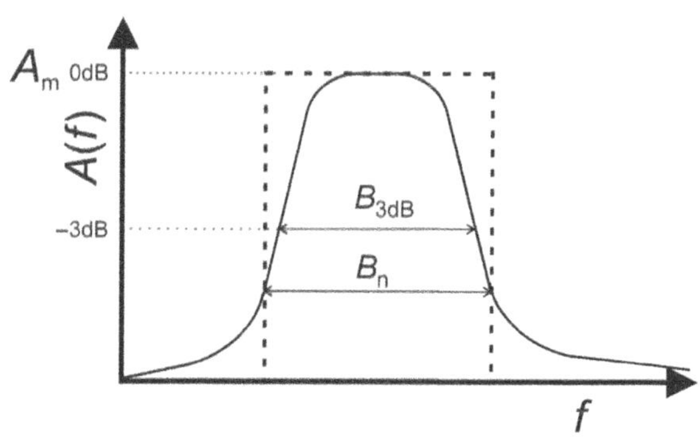

Figure 3.2 Defining the noise bandwidth B_n.

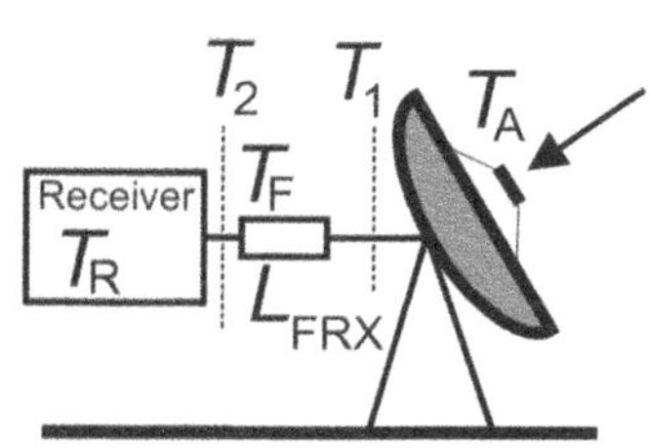

Figure 3.3 Receiving system.

2 The noise level is usually measured by a spectrum analyzer. In order to calculate the power spectral density N_0, the value of BW (bandwidth) is considered. The noise bandwidth of most spectrum analyzers is approximately equal to 1.2 times the bandwidth: $B_\mathrm{n} \approx 1.2\,\mathrm{BW}$.

The noise temperature T_1 of such a system is related to the reference plane of the antenna output connector. The noise temperature is given by the sum of noise temperatures T_A of a subsystem consisting of a feeder and a receiver. Using Eqs. (3.7) and (3.8), we get

$$T_1 = T_A + T_F(L_{FRX} - 1) + T_R L_{FRX}. \tag{3.10}$$

We obtain the noise temperature corresponding to the reference plane of the input connector of the receiver if T_1 is divided by the losses in the feeder L_{FRX}

$$T_2 = \frac{T_A}{L_{FRX}} + T_F\left(1 - \frac{1}{L_{FRX}}\right) + T_R = T_s. \tag{3.11}$$

The noise temperature T_2 represents an equivalent noise temperature T_S[3] of a receiving system, which equals to a temperature increase of internal conductance of the generator exciting the receiver, with respect to the reference temperature T_0. Such an increase is needed to achieve the same noise power at the output of the entire system when real elements are substituted by noise-free equivalents [3].

3.4 Figure of Merit G/T

The ratio between the gain of the antenna converted to the reference plane of the input connector of the receiver and the noise temperature of the system that is related to the same place is called the figure of merit of the receiving system

$$\frac{G}{T} = 10\log\left(\frac{G_{Rmax}}{L_R L_{FRX} L_{POL}} \cdot \frac{1}{T_s}\right) \text{ (dB/K)}. \tag{3.12}$$

Here, L_R are losses due to inaccurate orienting of the receiving antenna and L_{POL} are losses due to polarization mismatch.

The figure of merit G/T defined in this way can completely characterize the sensitivity of the receiving system. Smaller systems, e.g. for individual reception of TV signals from geostationary DBS (direct broadcasting satellite) satellites, have figures of merit of about 6 dB/K. Large ground satellite stations reach 35 dB/K. The DSN (Deep Space Network) 70 m telescope system achieves up to 60 dB/K in the X-band.

3 The input connector of the receiver (T_2) is the reference plane used to determine the noise temperature of the receiving system T_S. Here, we usually measure carrier-to-noise ratio C/N for the desired signal-to-noise ratio S/N at the output for analog signals, or E_b/N_0 (the ratio of one bit energy to spectral noise density) necessary to achieve the desired bit error ratio, BER, for digital signals.

References

1 Keysight Technologies. (2019). *Fundamentals of RF and Microwave Noise Figure Measurements*. Keysight Application Note 5952-8255. https://www.keysight.com/us/en/assets/7018-06808/application-notes/5952-8255.pdf (accessed 7 August 2024).

2 Keysight Technologies. (2014). *Noise Figure Measurement Accuracy – The Y-Factor Method*. Keysight Application Note 5952-3706E. https://www.ab4oj.com/test/docs/5952-3706E.pdf (accessed 7 August 2024).

3 Kasal, M. (2007). *Directional and Satellite Links* (in Czech). https://docplayer.cz/12896748-Smerove-a-druzicove-spoje.html (accessed 7 August 2024).

4

Communication Equation

An equation that relates the carrier power and the noise spectral density in a plane of the receiver connector can be used to characterize the energy budget of a microwave link between antennas with gains G_t and G_r (relative to an isotropic radiator) at a distance r [1]. Neglecting interferences and other disturbances of the link, the background noise of the signal can be represented by an additive white noise with a normal distribution AWGN (additive white Gaussian noise). In principle, we distinguish external noises received by an antenna and internal noises arising in a receiving device. Regardless of the noise source, the spectral density N_0 can be expressed as a product $k \cdot T_s$, where k is the Boltzmann constant and T_s is the equivalent noise temperature of the system. For the carrier power to noise spectral density ratio, we get

$$\frac{C}{N_0} = \frac{P_t G_t \frac{1}{L_0} G_r}{k T_s}. \tag{4.1}$$

Here, L_0 are free space propagation losses (see Chapter 5). If the concept of equivalent isotropic radiated power EIRP is introduced to characterize the entire transmission system, then

$$\mathrm{EIRP} = P_t G_t \tag{4.2}$$

Equation (4.1) can be written in the form

$$\frac{C}{N_0} = \mathrm{EIRP} \frac{1}{L_0} \frac{G_r}{T_s} \frac{1}{k} \tag{4.3}$$

The sensitivity of the receiving system is then completely characterized by the ratio G_r/T_s as mentioned above. The equation clearly shows contributions of a transmitting part, propagation losses, and a receiving part of the radio link to the

Reception of Weak Radio Signals from Space, First Edition. Miroslav Kasal.

overall energy budget. Since quantities significantly differ in size, a logarithmic form of the relation is usually used

$$\left(\frac{C}{N_0}\right)_{\text{dB}} = (\text{EIRP})_{\text{dB}} - (L_0)_{\text{dB}} + \left(\frac{G_r}{T_s}\right)_{\text{dB}} + 228.6. \tag{4.4}$$

The power of the carrier C and the equivalent isotropic radiated power EIRP are given in W, and the power spectral density of noise N_0 is given in W/Hz. $(L_0)_{\text{dB}}$ is the attenuation related to the propagation of electromagnetic waves in space. The figure of merit, G_r/T_s, is given in 1/K. Terms appearing in Equation (4.4) are expressed in a logarithmic scale, as emphasized by the subscript dB. The left-hand side of Equation (4.4) is therefore in dB · Hz, the ratio $(G_r/T_s)_{\text{dB}}$ in dB/K. The absolute term in Equation (4.4) is the reciprocal of the Boltzmann constant in dB.

If interested in the ratio of the carrier power to the total noise level in the transmission channel with noise bandwidth B_n (Hz), C/N can be calculated according to the formula

$$\left(\frac{C}{N}\right)_{\text{dB}} = \left(\frac{C}{N_0}\right)_{\text{dB}} - 10\log B_n \tag{4.5}$$

with $N = kT_sB_n$.

For data links with discrete modulations, we introduce the mean signal energy per transmitted bit E_b. For this energy related to the noise spectral density,

$$\frac{E_b}{N_0} = \frac{C}{N}\frac{B_n}{f_b}. \tag{4.6}$$

Here, f_b is the so-called information (bit) frequency (s^{-1}). Converting Eq. (4.6) into logarithmic form, we get

$$\left(\frac{E_b}{N_0}\right)_{\text{dB}} = (\text{EIRP})_{\text{dB}} - (L_0)_{\text{dB}} + \left(\frac{G_r}{T_s}\right)_{\text{dB}} + 228.6 - 10\log f_b. \tag{4.7}$$

Equation (4.7) applies to a two-level discrete modulation. In case of multi-level modulations (an M-bit symbol corresponds to one level), we get

$$\frac{E_s}{N_0} = \frac{E_b}{N_0}\log_2 M. \tag{4.8}$$

The E_b/N_0 ratio is of fundamental importance to express the signal-to-noise ratio per one bit of transmitted information. The ratio is related to the mean power of the received wanted signal, the information rate of binary symbols, and the spectral power density of the additive noise. Thanks to E_b/N_0, we can compare different discrete modulations from the viewpoint of energy demands on a certain error probability (see Fig. 4.1).

A low E_b/N_0 ratio is required in case of space links, and a poorer spectral efficiency f_b/B_n is the price to be paid for. In case of terrestrial microwave links,

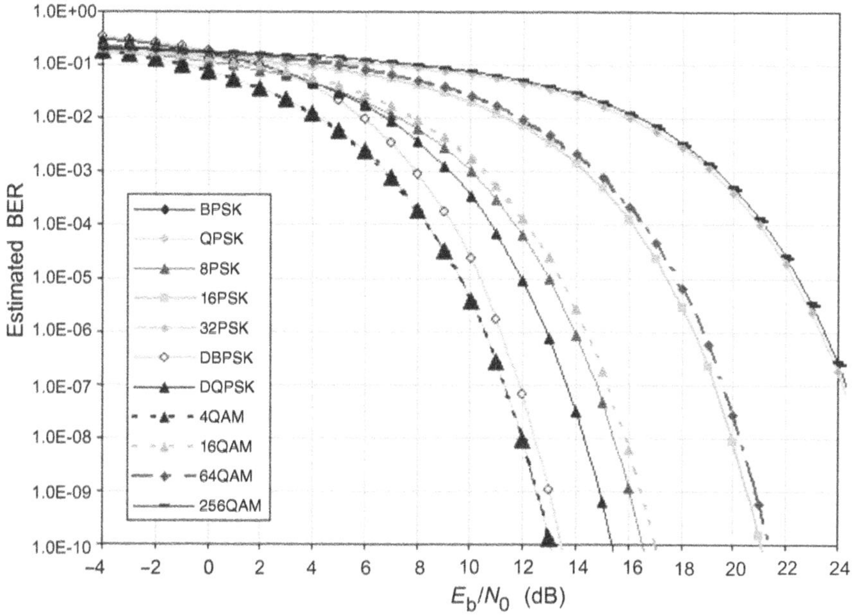

Figure 4.1 Theoretical dependence of the bit error rate, BER, on E_b/N_0 ratio for basic digital modulations. *Source: 2. Mathuranathan, V. (2010). Performance Comparison of Digital Modulation Techniques.* https://www.gaussianwaves.com/2010/04/performance-comparison-ofdigital-modulation-techniques-2 (accessed August 7 2024).

a higher spectral efficiency is usually preferred (a relatively larger E_b/N_0 ratio is required). Terrestrial radio relay links are usually designed for C/N about 30–40 dB above the threshold (without fading). Satellite links operate with a value of 8–10 dB (max. 16 dB in case of DBS satellites) above the threshold.

References

1 Kasal, M. (2007). *Directional and Satellite Links* (in Czech). https://docplayer.cz/12896748-Smerove-a-druzicove-spoje.html (accessed August 7 2024).

2 Mathuranathan, V. (2010). *Performance Comparison of Digital Modulation Techniques.* https://www.gaussianwaves.com/2010/04/performance-comparison-of-digital-modulation-techniques-2 (accessed August 7 2024).

5

Wave Propagation

5.1 Free Space Wave Propagation

Two isotropic radiators are considered at a distance r (see Fig. 5.1). The first radiator (at point 1) is fed by radio frequency signal of power P_1. What power P_2 is received by the second isotropic antenna at point 2? At all points on a sphere of radius r, the magnitude of the Poynting vector is

$$\Pi_r = \frac{P_1}{4\pi r^2}. \tag{5.1}$$

The power in the input resistance of the receiver perfectly matched to the isotropic radiator at point 2 is

$$P_2 = \Pi_r S_i. \tag{5.2}$$

Here, $S_i = \lambda^2(4\pi)^{-1}$ is the effective surface of the isotropic radiator Eq. (2.7). For the power transmission between points 1 and 2, we can write [1]

$$A_0 = \frac{P_2}{P_1} = \frac{\frac{\Pi_r \lambda^2}{4\pi}}{\Pi_r 4\pi r^2} = \left(\frac{\lambda}{4\pi r}\right)^2 = \frac{1}{L_0}. \tag{5.3}$$

The quantity[1]

$$L_0 = \left(\frac{4\pi r}{\lambda}\right)^2 \tag{5.4}$$

is called free space propagation loss of the forward electromagnetic wave. Usually, a logarithmic scale is applied, and the quantity is called attenuation of free space

1 Obviously, losses are directly proportional to the square of the distance and inversely proportional to the square of the wavelength. In other words, the losses at 10 GHz are 100x higher than at 1 GHz (by 20 dB). Let us consider that an antenna of the same size can have 20 dB higher gain at 10 GHz (2.14). If such antennas are at both ends of the link, the power budget is ultimately 20 dB more favorable at 10 GHz, but with significantly more directional antennas.

Reception of Weak Radio Signals from Space, First Edition. Miroslav Kasal.

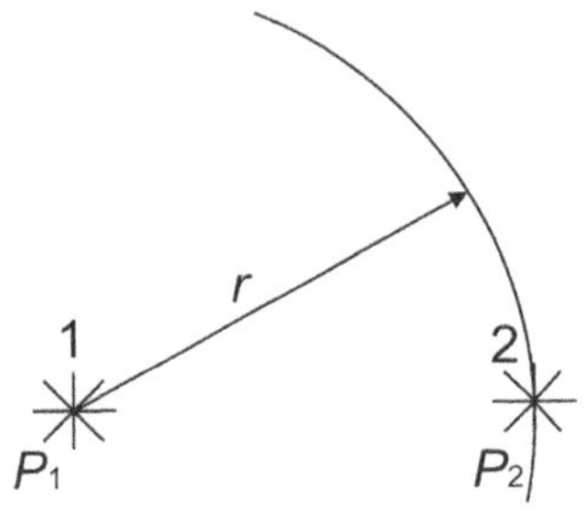

Figure 5.1 Two isotropic radiators at a distance *r*.

forward electromagnetic wave[2]

$$L_{0\mathrm{dB}} = 10 \log L_0. \tag{5.5}$$

5.2 Influence of the Atmosphere

The atmosphere is quite complex and variable from the viewpoint of propagation of microwave signals. An immediate meteorological situation can significantly influence the absorption fading of the signal. In the atmosphere, even other phenomena may also occur, such as scattering by hydrometeors and diffraction and refraction at the interface of air layers. All these phenomena are frequency selective, i.e. the effects at different frequencies are different. The highest absorption attenuation appears at the resonance of molecules of atmospheric gases – water vapor (H_2O) and oxygen (O_2). Figure 5.2 depicts absorption attenuation versus frequency.[3] Figure 5.3 shows the temperature dependency of atmospheric gases on the distance from Earth's surface [2].

5.2.1 Precipitation Cloud in Front of Antenna

The precipitation cloud in front of the receiving antenna acts as an attenuator at the receiver input. The signal is naturally attenuated due to the insertion loss of the cloud, and the noise temperature of the antenna is increased as well [3]. The situation is illustrated by Figure 5.4.

2 If the attenuation $L_{0\,\mathrm{dB}}$ thus calculated is substituted into Eqs. (4.4) and (4.7), we get the highest possible ratio C/N_0, or E_b/N_0, corresponding to an ideal, lossless propagation.
3 In Figure 5.2, the highlighted band around 60 GHz is affected by strong absorption due to oxygen molecules. Here the attenuation reaches more than 10 dB km^{-1}, which makes long-distance communication impossible. The use of this band for indoor communication and for spatial separation of adjacent "point-to-multipoint" networks is currently under consideration.

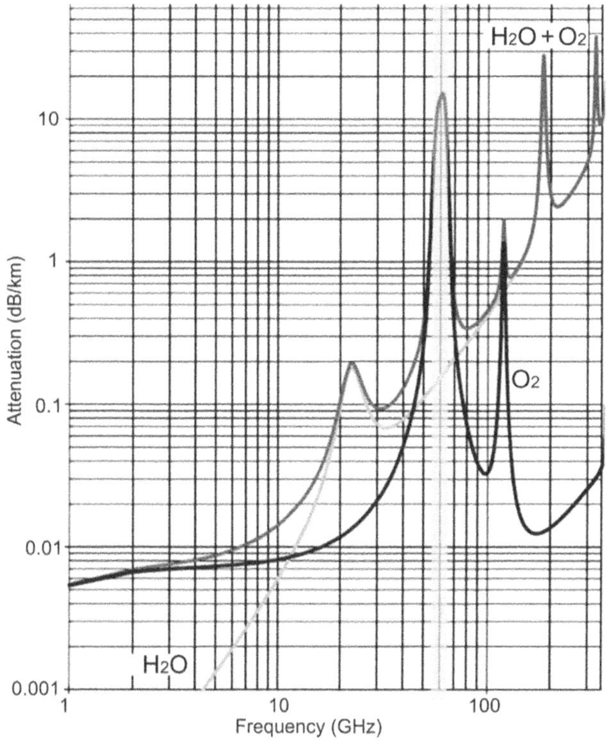

Figure 5.2 Atmospheric absorption attenuation as a function of frequency. The frequency dependencies of the attenuation due to water vapor and oxygen molecules are shown separately and in total.

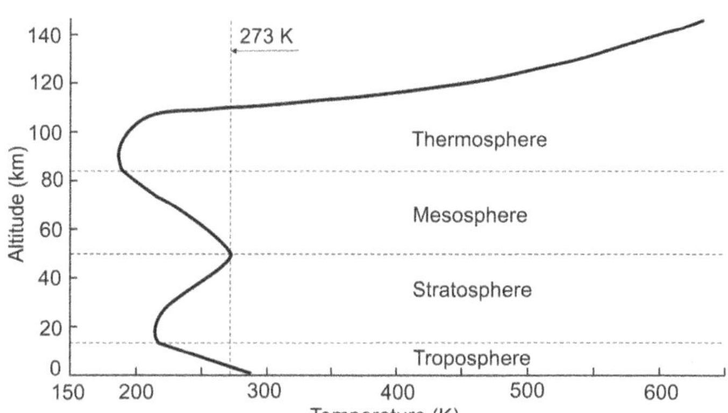

Figure 5.3 Temperature gradient in the standard atmosphere. *Source: 2. Strobel, N. (2022). Astronomy Notes.* https://www.astronomynotes.com/solarsys/s3c.html (accessed 7 August 2024).

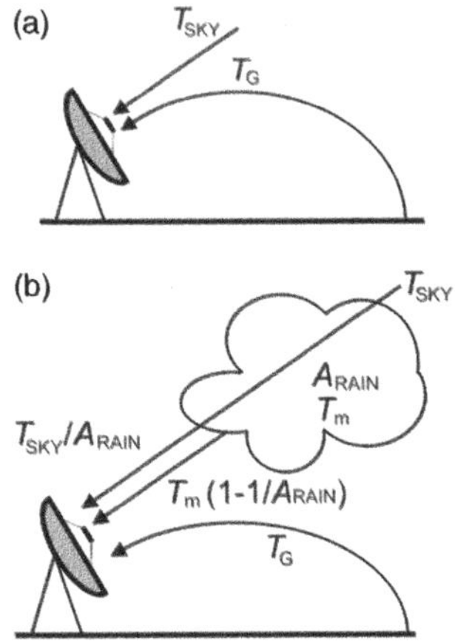

Figure 5.4 Influences related to the noise temperature of the antenna: (a) under clear sky; and (b) under rain.

If the antenna is oriented to clear sky (see Fig. 5.4a), its noise temperature T_A consists of two components

$$T_A = T_{SKY} + T_G. \tag{5.6}$$

Here, T_{SKY} is the brightness temperature of the sky in the radiation direction, and T_G is Earth's contribution received by the side lobes of the antenna radiation pattern. Let us assume a massive rain cloud in the main lobe direction of the antenna (see Fig. 5.4b), insertion loss A_{RAIN}, and thermodynamic temperature T_m. The expected signal attenuation is accompanied by the increase of noise temperature of the antenna under rain T_{AR}, which is equivalent to the behavior of the attenuator [4].[4]

$$T_{AR} = \frac{T_{SKY}}{A_{RAIN}} + T_m\left(1 - \frac{1}{A_{RAIN}}\right) + T_G. \tag{5.7}$$

5.2.2 Ionosphere Influence

The ionosphere is a part of the upper atmosphere that is partially ionized (plasma). Ultraviolet and corpuscular radiation from the Sun ionizes atoms and molecules in the ionosphere. The degree of ionization varies with time of the day, the season, geographic location, and the phase of the solar cycle. The ionosphere can be divided into several layers: D (60–90 km), E (90–150 km), and F (150–800 km).

The ionosphere is positively charged by about 10^6 C. The charge of Earth's surface is the same approximately, but the polarity is opposite. The voltage between Earth's surface and the ionosphere is about 360 kV and the current about 1 800 A during a clear day. At decameter waves, refraction in ionospheric

4 Let's compare Eqs. (5.7) and (3.11).

layers appears, which enables long-distance communication at these frequencies. The ionosphere influences higher frequencies as well. Nevertheless, refraction is less significant here.

Thanks to the interaction of Earth's magnetic field and electromagnetic wave in an ionized environment, the plane of linear polarization is varied (so-called Faraday rotation or Faraday effect).[5] Quantitatively, the phenomenon can be described by [5]

$$\Phi = \frac{2.36 \times 10^4}{f^2} \int N B \cos\theta_B \, dl. \tag{5.8}$$

Here, Φ denotes the rotation of the plane of linear polarization (radians), N is the number of electrons in volumetric unit of ionized layer (the number of electrons per m^3), B is the magnetic induction (T), and $B\cos\theta_B$ is the projection into the direction of propagation. The quantity

$$\mathrm{TEC} = \int N \, dl \tag{5.9}$$

is called "total electron content" and expresses the volume density of electrons integrated along the wave trajectory in the ionosphere, i.e. (the number of electrons per $m^{-3} \cdot m$).

Equation (5.8) shows that Faraday rotation is inversely proportional to the square of the frequency and is significant at low frequencies. Figure 5.5 shows

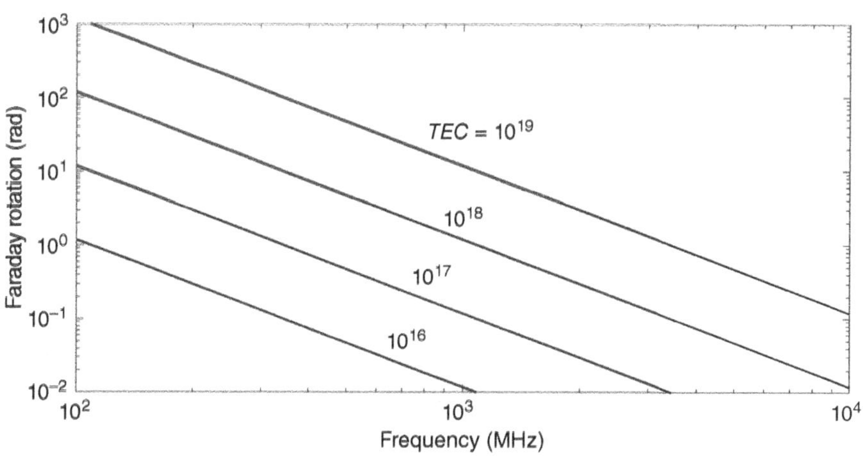

Figure 5.5 Frequency dependency of Faraday rotation for different values of TEC.
Source: 5. ALLNUTT, J.E. (1989). Satellite-to-Ground Radiowave Propagation: Theory, Practice, and System Impact at Frequencies above 1 GHz. London, UK: Peter Peregrinus.

5 Michael Faraday (1791–1867) discovered several fundamental phenomena in electromagnetism.

Eq. (5.8) graphically. At a quiet Sun, TEC ~ 10^{17} (average value; electrons m^{-2}). The plane of linear polarization is rotated 10 rad (!) at 100 MHz, 1 rad at 400 MHz, and 0.1 rad at 1 GHz. Above 5 GHz, the rotation is practically negligible. With increased solar activity, TEC ~ 10^{18} (electrons m^{-2}) and more. On the contrary, TEC is even less than 10^{16} (electrons m^{-2}) at the minimum of solar activity. The consequences are obvious. Faraday rotation can be best eliminated by using circular polarization, which is advantageous also due its insensitivity to multipath propagation.[6]

Refraction appears next to the rotation of the polarization plane during the wave propagation in the ionosphere, and this causes a trajectory extension, which can be estimated as follows

$$\Delta R = \frac{40.3}{f^2} \text{TEC (m; Hz, number of electrons m}^{-2}\text{)}. \tag{5.10}$$

This effect is particularly undesirable for satellite navigation signals. At $TEC = 10^{17}$ (electrons m^{-2}) and frequency 1 575 MHz, the trajectory extension is $\Delta R = 1.6$ m. At $TEC = 10^{18}$ (electrons m^{-2}), the extension equals 16 m.

5.3 Wave Propagation by Reflection

Conductive surfaces reflect waves at microwave frequencies, and the more conductive they are, the better. Considering the situation in Figure 5.6, the power transfer between points 1 and 2 can be written as [6]

$$A_{12} = \frac{1}{L_{12}} = \left(\frac{\lambda}{4\pi R_1}\right)^2 \left(\frac{S_{\text{eff}}}{S_{\text{i}}}\right) \left(\frac{\lambda}{4\pi R_2}\right)^2. \tag{5.11}$$

Here, $S_{\text{eff}} = \eta \cos\beta$ is the effective reflecting surface with geometric surface S and reflection coefficient η. S_{i} is the effective surface of the isotropic radiator.

Substituting into Eq. (5.11), we get

$$A_{12} = \frac{\eta S \cos\beta\, \lambda^2}{(4\pi)^3 (R_1 R_2)^2}. \tag{5.12}$$

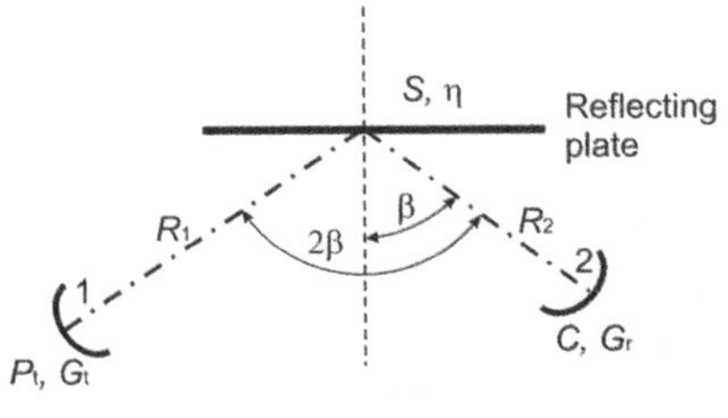

Figure 5.6 Passive reflective surface. *Source: 6. ĎUROVIČ, S. and WILFERT, O. (1991). Radio relay links (in Czech). In: Selected Scientific and Professional Theses of BUT: 5–71. Brno, CR: Brno University of Technology.*

6 The rotation of circular polarization is changed to the opposite one when the wave is reflected, and the reflected wave is greatly suppressed by the antenna.

If $R_1 = R_2 = D$ and $\beta = 0$, we get the so-called radar equation

$$A_\mathrm{r} = \frac{\eta S \lambda^2}{(4\pi)^3 D^4}. \tag{5.13}$$

References

1 Prokop, J. and Vokurka, J. (1980). *Electromagnetic Waves Propagation and Antennas* (in Czech). Prague, CR: SNTL/ALFA.

2 Strobel, N. (2022). *Astronomy Notes*. http://www.astronomynotes.com/solarsys/s3c.htm (accessed 7 August 2024).

3 Kasal, M. (2007). *Directional and Satellite Links* (in Czech). https://docplayer.cz/12896748-Smerove-a-druzicove-spoje.html (accessed 7 August 2024).

4 Maral, G. and Bousquet, M. (1993). *Satellite Communication Systems*. New York: Wiley.

5 Allnutt, J.E. (1989). *Satellite-to-Ground Radiowave Propagation: Theory, Practice, and System Impact at Frequencies Above 1 GHz*. London, UK: Peter Peregrinus.

6 Ďurovič, S. and Wilfert, O. (1991). Radio relay links (in Czech). In: *Selected Scientific and Professional Theses of BUT*: 5–71. Brno, CR: Brno University of Technology.

6

Measuring the Figure of Merit G/T of the Receiving System

Every design, and especially the implementation of a microwave receiving system, should be completed by measuring achieved parameters. Even during the design, the calculated and measured parameters of system sub-elements are inserted into the system model, considering corresponding weights. Based on this, resulting parameters are estimated. The figure of merit of the receiving system was defined in Chapter 3. In Chapter 4, the parameter was shown to describe system sensitivity completely. Usually, the figure of merit cannot be measured in laboratory conditions due to the physical dimensions of the antenna and the distance the reference signal source should be placed at, with the space between perfectly anechoic. To measure the G/T ratio and derived quantities, extraterrestrial noise sources (the Sun and the Moon primarily) are used [1–5]. For systems with a larger aperture, other space sources can be used (Cassiopeia A, Taurus, Sagittarius, etc.) [6, 7]. To make such measurements correspond to reality, the measurement procedure has to be carried out carefully and with knowledge of the context.

6.1 Y-Factor

If the feeder between the receiver and the antenna is a lossless one, the equivalent noise temperature of the system is, according to Eq. (3.11),

$$T_\mathrm{s} = T_\mathrm{A} + T_\mathrm{R}. \tag{6.1}$$

The method called Y-factor consists in measuring the ratio of radiation powers received from two different (well-defined) sources characterized by their temperatures T_{A1} and T_{A2}

$$Y = \frac{P_1}{P_2} = \frac{k(T_{\mathrm{A1}} + T_\mathrm{R})B_\mathrm{n}}{k(T_{\mathrm{A2}} + T_\mathrm{R})B_\mathrm{n}} = \frac{T_{\mathrm{A1}} + T_\mathrm{R}}{T_{\mathrm{A2}} + T_\mathrm{R}}. \tag{6.2}$$

Reception of Weak Radio Signals from Space, First Edition. Miroslav Kasal.

From there,

$$T_{\mathrm{R}} = \frac{T_{\mathrm{A1}} - YT_{\mathrm{A2}}}{Y - 1}. \tag{6.3}$$

As a reference level, we consider the power P_2. P_2 is measured with the antenna pointed at a clear sky. Since all measurements are performed at the same noise bandwidth B_{n} (Hz), the power spectral density (W/Hz) is used (for sake of simplicity) instead of power (W): $P_2/B_{\mathrm{n}} = P_{\mathrm{SKY}}$. Therefore,

$$P_{\mathrm{SKY}} = k(T_{\mathrm{A2}} + T_{\mathrm{R}}) = kT_{\mathrm{s}}. \tag{6.4}$$

Then, the antenna is directed toward a black body in the sky (approximately), and power P_1 is measured as follows

$$P_1 = kT_{\mathrm{s}} + \frac{1}{2}\int_{\Omega} J_\nu S_{\mathrm{eff}}\, \mathrm{d}\Omega. \tag{6.5}$$

For the derivation of the second term of the right-hand side of Eq. (6.5), see Eq. (2.5). Substituting for S_{eff} (2.8) and integrating over the solid angle, we get

$$P_1 = kT_{\mathrm{s}} + \frac{1}{2}\,\frac{2kT_{\mathrm{A1}}}{\lambda^2}\,\frac{\lambda^2 G}{4\pi}. \tag{6.6}$$

For Y, we can write

$$Y = \frac{P_1}{P_2} = \frac{kT_{\mathrm{s}} + \frac{1}{2}\,\frac{2kT_{\mathrm{A1}}}{\lambda^2}\,\frac{\lambda^2 G}{4\pi}}{kT_{\mathrm{s}}} = 1 + \frac{kT_{\mathrm{A1}}G}{4\pi kT_{\mathrm{s}}}. \tag{6.7}$$

Using the relation (2.4), we get

$$Y = 1 + \frac{\frac{G}{T_{\mathrm{s}}}\lambda^2 J_\nu}{8\pi k}. \tag{6.8}$$

Finally, the expression for the figure of merit of the receiving system G/T (1/K) at the angular width of the radiating body θ_{bb} (sr) is

$$\frac{G}{T_{\mathrm{s}}} = \frac{8\pi k(Y - 1)}{\lambda^2 N_\nu}. \tag{6.9}$$

If the power spectral density of a selected black-body radiation is known at the operating frequency N_ν, the figure of merit G/T can be determined considering the ratio Y. The accuracy of the method is greater the larger Y and the more accurately N_ν is known. However, plenty of other circumstances affect result accuracy. Mainly, the derivation assumes uniformity of the antenna beam and uniformity of the body and sky radiation (the parameters do not depend on the angles θ and φ). Since these assumptions are not generally valid (especially for the elevation angle θ), both angles θ and φ have to be reported together with the value of Y. The beam of a highly directional antenna is of a Gaussian-like shape. The crucial role is played by the ratio between the antenna beam width HPBW $= \theta_{3\,\mathrm{dB}}$ and the angular width of the black body θ_{bb}. Losses due to atmospheric propagation were neglected. Therefore, Eq. (6.9) is completed by two correction factors K_1 and K_2 [1]

$$\frac{G}{T_s} = \frac{8\pi k(Y-1)}{\lambda^2 N_\nu} K_1 K_2. \tag{6.10}$$

K_1 is the factor correcting losses in the atmosphere. Choosing an appropriate time of measurements (a cloudless day with low humidity of the atmosphere), the losses can be usually reduced so much that K_1 can be set as $K_1 = 1$.[1] The factor K_2 corrects the spatial widening of radiation of the source (relative to the width of the antenna beam). Several methods have been developed to describe this correction [1, 8]. If an angular width of the black body θ_{bb} is higher than one fifth of the beam width $\theta_{3\,dB}$, the relation for K_2 is derived when approximating the antenna beam by the normalized Gaussian function (FWHM = $\theta_{3\,dB}$)

$$K_2 = \frac{\ln(2)\cdot\left(\dfrac{\theta_{bb}}{\theta_{3\,dB}}\right)^2}{1-\exp\left[-\ln(2)\cdot\left(\dfrac{\theta_{bb}}{\theta_{3\,dB}}\right)^2\right]}. \tag{6.11}$$

Since the angular width of the Moon and the Sun is ~0.5°, K_2 according to Eq. (6.11) can be used for microwave antennas up to a beam width of 2.5° (see Fig. 6.1).

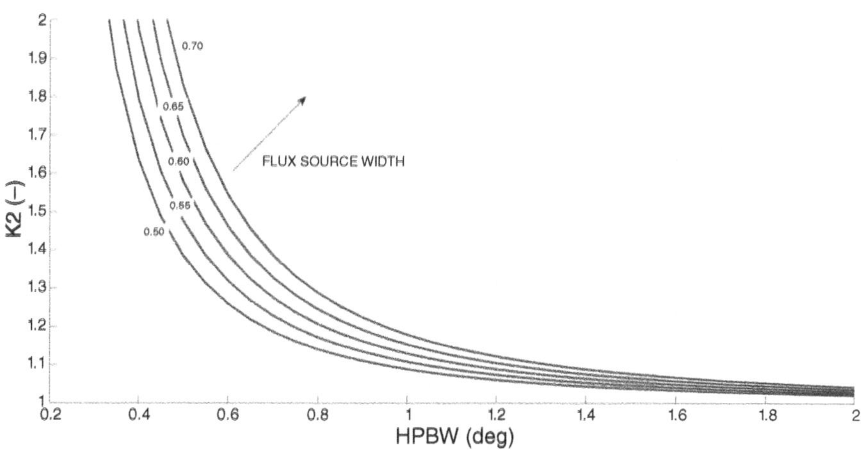

Figure 6.1 Dependence of the correction factor K_2 on HPBW of the antenna for different angular widths of the radiation source according to Eq. (6.11).

1 Of course, this statement does not apply in general. For example in the Ka band, where the absorption is increased by water molecules (see Fig. 5.2), the losses in the atmosphere cannot be usually neglected.

6.2 Measuring *G/T* Using the Moon

The Moon is probably the most suitable radiation source for measuring the G/T ratio of medium-sized systems (with an aperture of several meters in diameter). The Moon radiation is quite precisely defined, and achievable Y values are high enough for measuring such systems.

For the power spectral density of the Moon radiation, we get [1] with reference to Eqs. (2.2) and (2.4)

$$N_\nu = \frac{2kT_{\mathrm{Mo}}}{\lambda^2}\Omega_{\mathrm{Mo}}\ (\mathrm{W\,m^{-2}\,Hz^{-1}};\mathrm{K,m,sr}). \tag{6.12}$$

Here, T_{Mo} is the brightness temperature of the Moon and Ω_{Mo} is the solid angle of the Moon in steradians. The brightness temperature of the Moon slightly varies around the temperature $\overline{T}_0$ according to

$$T_{\mathrm{Mo}} = \overline{T}_0\left[1 - \frac{\overline{T}_1}{\overline{T}_0}\cos(\phi - \psi)\right]\ (\mathrm{K},\ ^\circ) \tag{6.13}$$

with

$$\begin{aligned}\overline{T}_0 &= 207.7 + \frac{24.43}{f}\ (\mathrm{K;GHz})\\ \frac{\overline{T}_1}{\overline{T}_0} &= 0.004212\cdot f^{1.224}\ (\mathrm{K,K;GHz})\\ \psi &= \frac{43.83}{1+0.0109\cdot f}\ (^\circ;\mathrm{GHz}).\end{aligned} \tag{6.14}$$

The quantities in Eq. (6.14) result from the expansion of the brightness temperature frequency dependence $T_{\mathrm{Mo}}(f)$, where $\overline{T}_0$ is the mean value and $\overline{T}_1$ the first harmonics. In general, the brightness temperature of the Moon depends on the solar mean anomaly (the position of the Earth-Moon system relative to the Sun during the year usually can be ignored) and the phase of the Moon ϕ Eq. (6.13). The angle ψ is then the phase delay.

When measuring and calculating, the influence of separate parameters on the accuracy should be considered. For a frequency 10.368 GHz, $\overline{T}_0 = 210$ K and the amplitude of brightness temperature variations during the monthly period is $\overline{T}_1 = 15.5$ K. Hence, the ratio of the highest to the lowest brightness temperature is $225.5/194.5 = 1.16$, which corresponds to the maximum inaccuracy ±0.32 dB. Such an inaccuracy can be neglected or not. This depends on the accuracy of other parameters, the Y ratio, and the beam angular width $\theta_{3\,\mathrm{dB}}$ especially. Similarly, the angular size of the Moon slightly varies from 0.48° to 0.56°. In most cases, we can consider the value $\Theta_{\mathrm{Mo}} = 0.5°$. To substitute into Eq. (6.12), the angular width of the black body used (here the Moon) Θ_{bb} has to be recalculated to the solid angle Ω_{bb}

$$\Omega_{bb} = 2\pi\left(1 - \cos\frac{\Theta_{bb}}{2}\right) \quad (\text{sr; }^\circ). \tag{6.15}$$

For the Moon, we can get $\Omega_{Mo} = 2\pi[1 - \cos(0.5/2)] = 5.98 \times 10^{-5}$ sr.

Example 6.1 The G/T ratio and the equivalent noise temperature T_S of a system with an offset reflector (aperture diameter 1.8 m) are investigated at a frequency 10 368 MHz.

3 June 2020, 21:05 UTC,

Moon (Moon Noise): $\phi = 160.7°$, azimuth Az = 179.2°, elevation El = 25.5° from the position 49°51′15″N and 16°22′30″E;
Clear Sky: Az = 179.2°, El = 35.5°.

By repeated measurements [9], we can obtain (see Fig. 6.2):

$$Y_{dB} = 0.95 \text{ dB},^2$$
$$Y = 1.24.$$

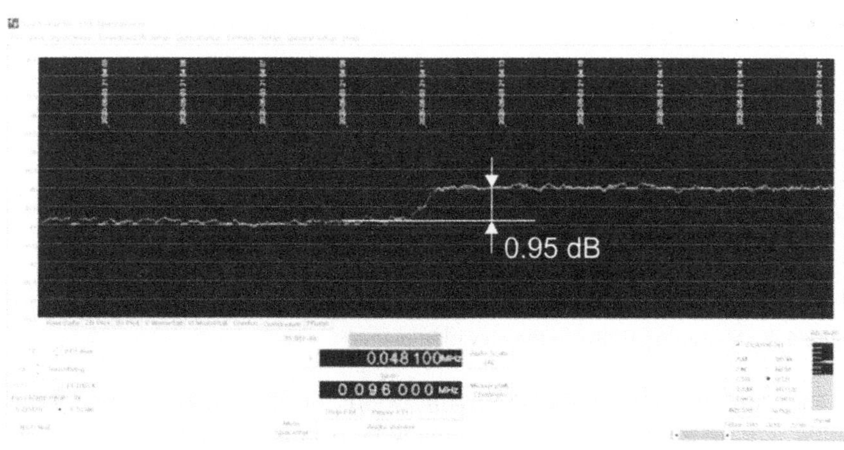

Figure 6.2 Measuring Y ratio as MN/CS.

2 Accurate measurement of small variations of a highly amplified noise is quite difficult. The program SpectraVue in the Continuum mode [9] is a suitable tool.

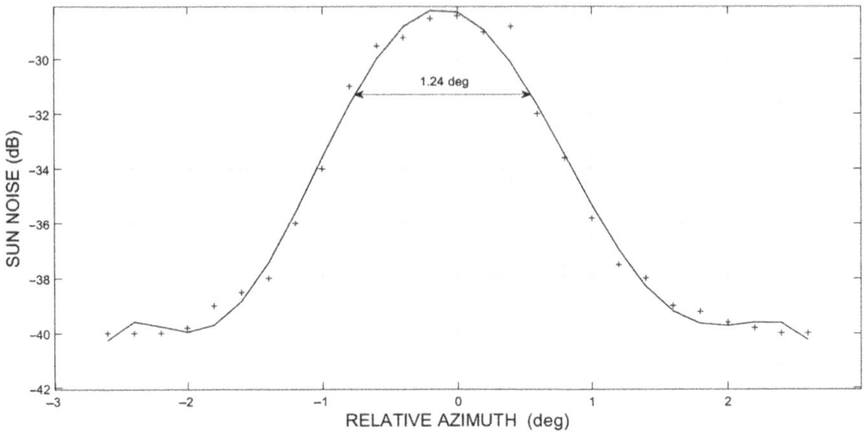

Figure 6.3 Measuring HPBW in the azimuth plane.

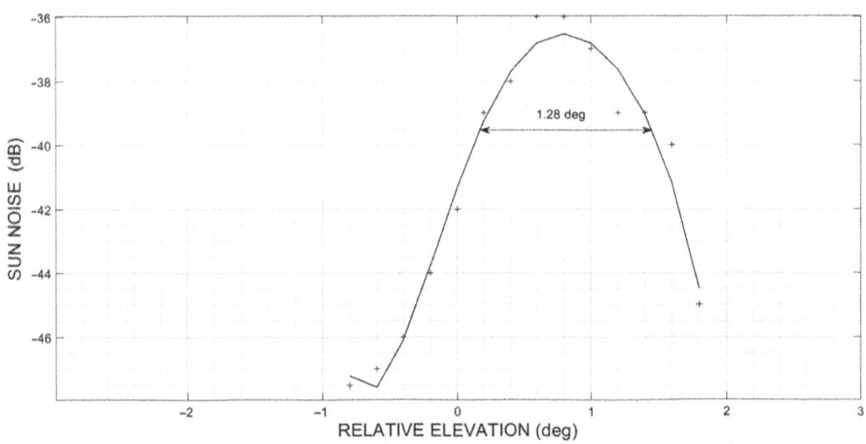

Figure 6.4 Measuring HPBW in the elevation plane.

Measuring the Sun noise (SN), the width of the main antenna beam HPBW = $\theta_{3\,\text{dB}}$ is obtained in two mutually perpendicular planes (see Figs. 6.3 and 6.4). In both the cases, SN was measured with a step 0.2°, and obtained values were interpolated with polynomials of the 7th and 5th order, respectively.[3]

The resulting value of HPBW is the arithmetic mean of both the measurements

$$\Theta_{3\,\text{dB}} = \frac{\Theta_{3\,\text{dBH}} + \Theta_{3\,\text{dBV}}}{2} = \frac{1.24 + 1.28}{2} = 1.26°$$

3 The measured values $\Theta_{3\,\text{dBH}}$ and $\Theta_{3\,\text{dBV}}$ show that the antenna beam is nearly rotationally symmetric.

From Eq. (6.14), the phase delay of the Moon at the frequency of 10.368 GHz is calculated

$$\psi = \frac{43.83}{1 + 0.0109 \cdot 10.368} = 39.4^\circ$$

and

$$\frac{\overline{T_1}}{\overline{T_0}} = 0.004212 \cdot (10.368)^{1.224} = 0.74.$$

For the brightness temperature of the Moon at the time of measurement Eq. (6.13), we get after substituting

$$T_{\text{Mo}} = 210[1 - 0.074\cos(160.7 - 39.4)] = 218\,\text{K}.$$

Finally, the power density of the received radiation of the Moon Eq. (6.12)

$$N_\nu = \frac{2 \cdot 1.38 \cdot 10^{-23} \cdot 218}{0.029^2} 5.98 \cdot 10^{-5} = 4.28 \cdot 10^{-22}\ \text{Wm}^{-2}\,\text{Hz}^{-1}.$$

Next, we calculate K_2 Eq. (6.11)

$$K_2 = \frac{\ln(2) \cdot \left(\frac{\Theta_{\text{Mo}}}{\Theta_{3\,\text{dB}}}\right)^2}{1 - \exp\left[-\ln(2) \cdot \left(\frac{\Theta_{\text{Mo}}}{\Theta_{3\,\text{dB}}}\right)^2\right]} = \frac{0.69 \cdot \left(\frac{0.5}{1.26}\right)^2}{1 - \exp\left[-0.69 \cdot \left(\frac{0.5}{1.26}\right)^2\right]} = 1.055.$$

Now, all quantities to be substituted into Eq. (6.10) are available

$$\frac{G}{T_s} = \frac{8 \cdot 3.14 \cdot 1.38 \cdot 10^{-23}(1.24 - 1)}{0.029^2 \cdot 4.28 \cdot 10^{-22}} \cdot 1 \cdot 1.055 = 231.25\,\text{K}^{-1}$$

or

$$\left(\frac{G}{T_s}\right)_{\text{dB}} = 10\ \log 231.25 = 23.6\,\text{dB}\,\text{K}^{-1}.$$

In Chapter 2, a relation for maximum directivity was presented as a function of HPBW (2.18). The relation can be modified as follows

$$G_{\max\ \text{dB}} = 10\log\left(\frac{35\,000}{\Theta_{3\,\text{dB}}^2}\right) = 10\log\left(\frac{35\,000}{1.26^2}\right) = 43.4\,\text{dBi}.$$

The equivalent noise temperature of the system can be calculated

$$T_s = 10^{\frac{G_{\max\ \text{dB}} - \left(\frac{G}{T_s}\right)_{\text{dB}}}{10}} = 10^{\frac{43.4 - 23.6}{10}} = 95.5\,\text{K}.$$

This is a very good value considering that the equivalent noise temperature of the low-noise amplifier (LNA) itself is around 60 K (noise figure $F_{\text{dB}} = 0.8\,\text{dB}$).

Using Eq. (2.15), the aperture efficiency can be further calculated

$$\eta = \left(\frac{\lambda}{\pi d}\right)^2 10^{\frac{G_{\max\,\mathrm{dB}}}{10}}. \tag{6.16}$$

For $d = 1.80$ m, $\lambda = 0.029$ m, and $G_{\max\,\mathrm{dB}} = 43.4\ \mathrm{dB_i}$, we get

$$\eta = \left(\frac{0.029}{3.14 \cdot 1.80}\right)^2 10^{\frac{43.4}{10}} = 0.58.$$

6.3 Measuring *G/T* Using the Sun

Compared to the Moon, the radiation of the Sun is much more intensive. Therefore, the Y values are higher and more accurate. Using the Sun, even small systems with HPBW < 2.5° can be measured. On the other hand, time variations of Sun's radiation are stronger (both within the solar cycle, and within the season or momentary eruptions). Since Sun's radiation (SF: solar flux) [10] is monitored by a number of observatories around the world,[4] current values of the intensity can be obtained even in second intervals at several frequencies. As described in Chapter 2, the solar radiation intensity is given in units of SFU = $10^{-22}\ \mathrm{W\,m^{-2}\,Hz^{-1}} = 10\,000$ Jy. E.g. data downloaded from the Learmonth Observatory in Western Australia contain measured SFU values at second intervals (see Table 6.1).

Example 6.2 Using the Solar Flux, measurements and calculations similar to the previous example will be performed.

7 June 2020 16:20 UTC

Sun: Az = 279.5°, El = 22.3° from the position 49°51′15″N and 16°22′30″E
Clear Sky: Az = 279.5°, El = 32.3°

The dependency in Figure 6.5 shows the transition between two levels. The higher level corresponds to SN and the lower one to CS. The difference is Y: $Y_{\mathrm{dB}} = 11.5$ dB, $Y = 14.1$.

Even if the data in Table 6.1 are for a different time (the observatory is in Australia), the Sun was calm that day, and SFU values varied slightly. SFU values can be therefore assumed (with high probability) to be the same at the time of our measurement. However, the frequency dependence of SFU is important. The fundamental wavelength of solar activity monitoring is $\lambda = 10.7$ cm.[5] If the

4 E.g. daily SFU values can be obtained at https://www.spaceweather.gc.ca/forecast-prevision/solar-solaire/solarflux/sx-4-en.php.
5 The wavelength 10.7 cm corresponds to a frequency of 2 804 MHz (in a vacuum).

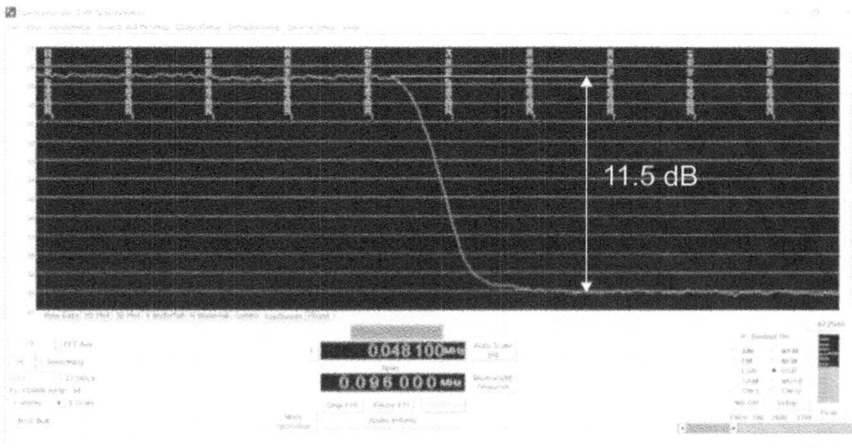

Figure 6.5 Measuring *Y* ratio as SN/CS.

Table 6.1 Sample of SFU values from 7 June 2020 at several frequencies, Learmonth Observatory in Western Australia.

Time	245	410	510	1415	2695	4995	8800	15400	[MHz]
092842	000015	000026	000033	000048	000062	000090	000146	000416	
092843	000015	000026	000033	000047	000061	000089	000148	000418	
092844	000014	000026	000032	000047	000063	000091	000148	000418	
092845	000014	000025	000033	000047	000062	000089	000148	000412	
092846	000014	000026	000033	000047	000061	000088	000147	000410	
092847	000014	000025	000033	000049	000061	000090	000146	000403	[SFU]
092848	000014	000025	000032	000049	000060	000091	000145	000400	
092849	000014	000025	000033	000048	000059	000088	000142	000399	
092850	000014	000025	000032	000047	000061	000088	000141	000398	
092851	000014	000025	000032	000047	000062	000087	000142	000401	
092852	000014	000025	000032	000047	000059	000084	000143	000399	

line for the time 09:28:44 (see Fig. 6.6) is plotted (for example), we can see that SF = 63 SFU at 2695 MHz, and SF = 212 SFU at 10 368 MHz (linear interpolation). The authors in [11] show another interpolation for SF_f between two adjacent frequencies f_1 and f_2

$$\mathrm{SF}_f = \left(\frac{\mathrm{SF}_{f_1}}{\mathrm{SF}_{f_2}}\right)^{\alpha_1} \cdot \mathrm{SF}_{f_2} \tag{6.17}$$

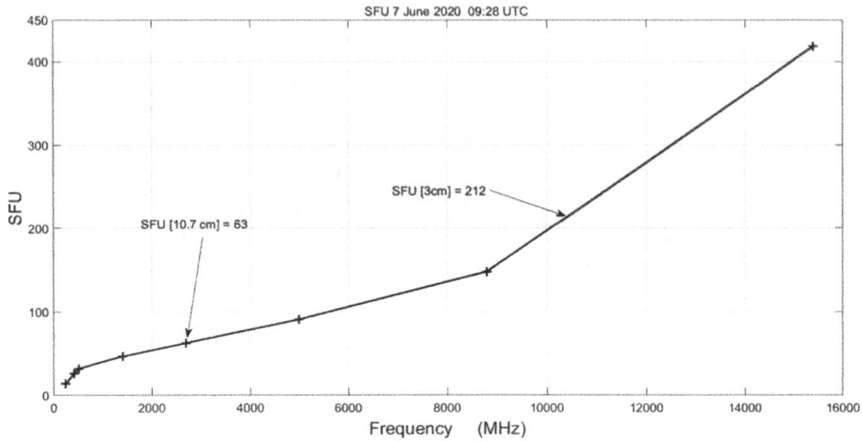

Figure 6.6 Dependence of SFU on frequency.

where f is the considered frequency and

$$\alpha_1 = \frac{\log(f/f_2)}{\log(f_1/f_2)}. \tag{6.18}$$

Substituting values from the same row of the table to Eqs. (6.18) and (6.17), we get $f = 10.368$ GHz

$$\alpha_1 = \frac{\log\left(\frac{10.368}{15.400}\right)}{\log\left(\frac{8.800}{15.400}\right)} = 0.707 \text{ and } \mathrm{SF}_{10.368} = \left(\frac{148}{418}\right)^{0.707} \cdot 418 = 200.6\,\mathrm{SFU}.$$

For the next calculation, a less optimistic value is taken from the chart: 212 SFU.

The angular width of the Sun also slightly varies depending on the frequency [1]

$$\Theta_S = 0.533 \cdot 10^{\frac{1}{\alpha_2}}. \tag{6.19}$$

Here for the frequency f in GHz

$$\alpha_2 = \frac{f^4}{1818} - \frac{f^3}{23.35} + 1.22f^2 - 2.66f + 11.43. \tag{6.20}$$

Substituting into the last two equations, we get $f = 10.368$ GHz

$$\alpha_2 = \frac{10.368^4}{1818} - \frac{10.368^3}{23.35} + 1.22 \cdot 10.368^2 - 2.66 \cdot 10.368 + 11.43 = 74$$

and

$$\Theta_S = 0.533 \cdot 10^{\frac{1}{74}} = 0.55°.$$

Next, the correction factor K_2 is calculated

$$K_2 = \frac{\ln(2) \cdot \left(\frac{0.55}{1.26}\right)^2}{1 - \exp\left[-\ln(2) \cdot \left(\frac{0.55}{1.26}\right)^2\right]} = 1.067.$$

Equation (6.10) yields

$$\frac{G}{T_s} = \frac{8 \cdot 3.14 \cdot 1.38 \cdot 10^{-23}(14.1 - 1)}{0.029^2 \cdot 212 \cdot 10^{-22}} \cdot 1 \cdot 1.067 = 272\,\mathrm{K}^{-1}$$

and in a logarithmic scale

$$\left(\frac{G}{T_s}\right)_{\mathrm{dB}} = 10\log 272 = 24.3\,\mathrm{dB\,K}^{-1}.$$

The difference, compared to the measurement of the Moon, is $0.7\,\mathrm{dB\,K}^{-1}$. The value is very acceptable, considering the measurements not being performed at the same time or on the same day (under the same atmospheric conditions). Assuming $K_1 = 1$, this effect was neglected in both cases. No matter that the elevation was similar in both cases, the measurement location of the sky radiation was different. The last relation shows

$$T_s = 10^{\frac{43.4-24.3}{10}} = 81.3\,\mathrm{K},$$

which is a very good value ($T_{\mathrm{LNA}} = 60$ K). Both experiments show that the antenna has a slightly lower aperture efficiency and an excellent noise temperature. Both properties are achieved thanks to an offset reflector.

Example 6.3 At the observatory in Panská Ves (50.5278°N, 14.5651°E), the Y system with a 10 m rotationally symmetrical parabolic antenna was measured at frequency 2 320 MHz on 11 August 2010. The directivity pattern, which was obtained by measuring the solar noise in elevation and by interpolating with a 12th-order polynomial, is shown in Figure 6.7. Furthermore, the antenna directivity pattern is assumed being rotationally symmetrical.

During measurements, $\mathrm{SF}_{10.7\,\mathrm{cm}} = 85.8$ SFU. Due to the proximity of both frequencies, the value is taken as a reference. Next, from the chart, the ratio $Y_{\mathrm{dB}} = 19.8$ dB, $Y = 95.5$ was taken.

For the angular width of the Sun, we get

$$\alpha_2 = \frac{2.320^4}{1818} - \frac{2.320^3}{23.35} + 1.22 \cdot 2.320^2 - 2.66 \cdot 2.320 + 11.43 = 11.3,$$

$$\Theta_S = 0.533 \cdot 10^{\frac{1}{11.3}} = 0.65°.$$

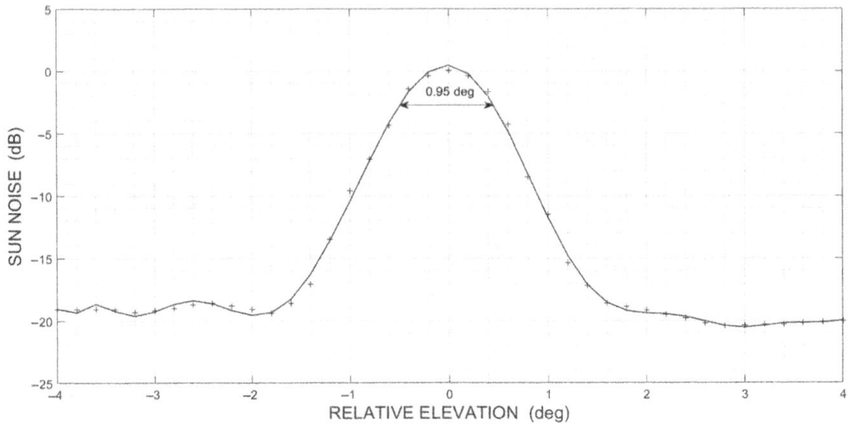

Figure 6.7 Radiation pattern of the 10 m Y parabola in the elevation plane.

Similarly to the previous case, we calculate

$$K_2 = \frac{\ln(2) \cdot \left(\frac{0.65}{0.95}\right)^2}{1 - \exp\left[-\ln(2) \cdot \left(\frac{0.65}{0.95}\right)^2\right]} = 1.17,$$

$$\frac{G}{T_s} = \frac{8 \cdot 3.14 \cdot 1.38 \cdot 10^{-23}(95.5 - 1)}{0.129^2 \cdot 85.8 \cdot 10^{-22}} \cdot 1 \cdot 1.17 = 268.6\,\mathrm{K}^{-1},$$

$$\left(\frac{G}{T_s}\right)_{\mathrm{dB}} = 10\log 268.6 = 24.3\,\mathrm{dB\,K}^{-1},$$

$$G_{\max\ \mathrm{dB}} = 10\log\left(\frac{35\,000}{\Theta^2_{3\,\mathrm{dB}}}\right) = 10\log\left(\frac{35\,000}{0.95^2}\right) = 45.9\,\mathrm{dBi},$$

$$T_s = 10^{\frac{45.9-24.3}{10}} = 144.5\,\mathrm{K}.$$

Finally, we evaluate the aperture efficiency Eq. (6.16) for $d = 10$ m, $\lambda = 0.129$ m and $G_{\max\ \mathrm{dB}} = 45.9$ dBi

$$\eta = \left(\frac{0.129}{3.14 \cdot 10}\right)^2 10^{\frac{45.9}{10}} = 0.65.$$

A low noise temperature T_s and a high aperture efficiency η indicates a very well-adjusted system.

Example 6.4 Under similar conditions, the second Z system with a 10 m central reflector was measured at the observatory in Panská Ves at a frequency of 2 320 MHz. The directivity pattern, which was obtained by measuring the solar

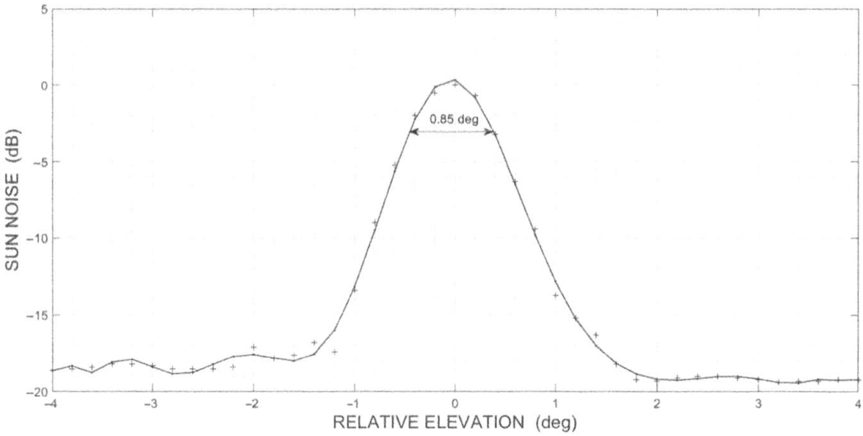

Figure 6.8 Radiation pattern of the 10 m parabola in the elevation plane.

noise in elevation and by interpolating with a 12th-order polynomial, is shown in Figure 6.8. Furthermore, the antenna directivity pattern is assumed being rotationally symmetrical. From the chart, we obtained the ratio

$$Y_{\mathrm{dB}} = 19.0 \quad \mathrm{dB}, Y = 79.4.$$

From the chart in Figure 6.8, we get HPBW

$$\Theta_{3\,\mathrm{dB}} = 0.85^\circ.$$

The angular width of the Sun is the same as in the previous case

$$\Theta_{\mathrm{S}} = 0.65^\circ.$$

Similarly, we calculate

$$K_2 = \frac{\ln(2) \cdot \left(\frac{0.65}{0.85}\right)^2}{1 - \exp\left[-\ln(2) \cdot \left(\frac{0.65}{0.85}\right)^2\right]} = 1.22,$$

$$\frac{G}{T_{\mathrm{s}}} = \frac{8 \cdot 3.14 \cdot 1.38 \cdot 10^{-23}(79.4 - 1)}{0.129^2 \cdot 85.8 \cdot 10^{-22}} \cdot 1 \cdot 1.22 = 232.3\ \mathrm{K}^{-1},$$

$$\left(\frac{G}{T_{\mathrm{s}}}\right)_{\mathrm{dB}} = 10 \log 232.3 = 23.7\,\mathrm{dB\,K}^{-1},$$

$$G_{\max\ \mathrm{dB}} = 10 \log\left(\frac{35\,000}{\Theta^2_{3\,\mathrm{dB}}}\right) = 10 \log\left(\frac{35\,000}{0.85^2}\right) = 46.8\,\mathrm{dBi},$$

$$T_{\mathrm{s}} = 10^{\frac{46.8-23.7}{10}} = 204.2\,\mathrm{K}.$$

Finally, we can evaluate the aperture efficiency Eq. (6.16) for $d = 10\,\text{m}$, $\lambda = 0.129$ m and $G_{\max\,\text{dB}} = 46.8$ dBi

$$\eta = \left(\frac{0.129}{3.14 \cdot 10}\right)^2 10^{\frac{46.8}{10}} = 0.81\cdot$$

The aperture efficiency η is very high. However, the noise temperature T_s is higher. Both of these facts are probably caused by the spillover of the reflector. Even if the antenna gain is higher, the figure of merit $(G/T_s)_{\text{dB}}$ is for 0.6 dB K^{-1} lower compared to the Y system.

Example 6.5 At the observatory of the Institute of Atmospheric Physics in Panská Ves, even a system with a 4.2 m reflector was measured at a frequency of 8.4 GHz. Sun measurements, which were performed on 26 August 2020 at the elevation about 40°, provided the following results: $Y_{\text{dB}} = 15$ dB and beam widths in both planes (see Figs. 6.10 and 6.11). Current solar flux (SFU) data were obtained from Learmonth Observatory: 092108 000015 000027 000039 000052 000068 000107 000210 000507 (see Fig. 6.9 for frequencies as in Table 6.1). At the standard value of SF (2 695 MHz) = 68 SFU, we obtained $N_\nu = 200 \times 10^{-22}\,\text{Wm}^{-2}\text{Hz}^{-1}$ at a frequency of 8.4 GHz.

The beam width is determined as an arithmetic mean (Figs. 6.10 and 6.11),

$$\Theta_{3\,\text{dB}} = \frac{\Theta_{3\,\text{dBH}} + \Theta_{3\,\text{dBV}}}{2} = \frac{0.56 + 0.66}{2} = 0.61°,$$

$$\alpha_2 = \frac{8.4^4}{1818} - \frac{8.4^3}{23.35} + 1.22 \cdot 8.4^2 - 2.66 \cdot 8.4 + 11.43 = 52.5,$$

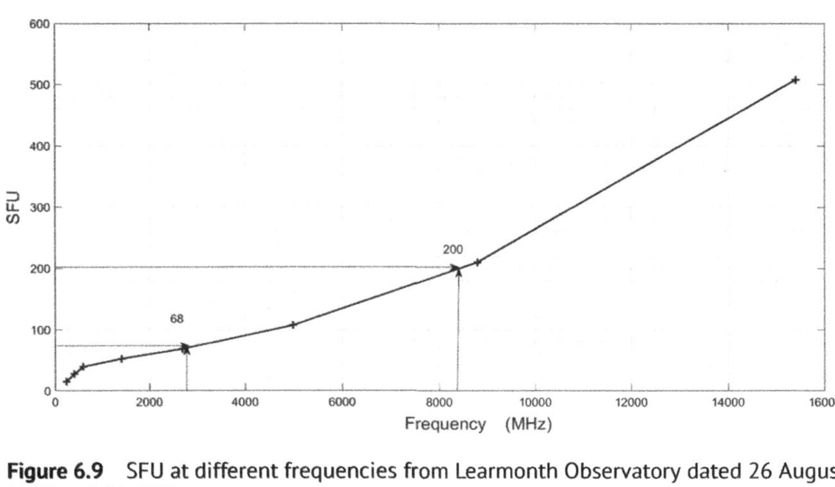

Figure 6.9 SFU at different frequencies from Learmonth Observatory dated 26 August 2020 at 9:21 UTC.

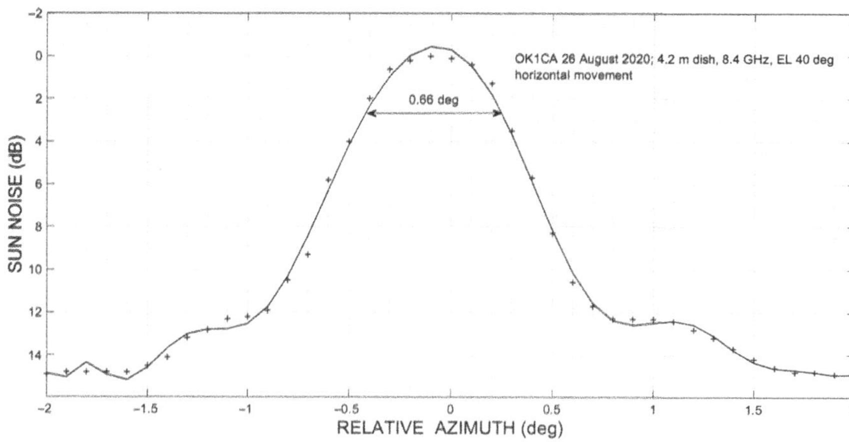

Figure 6.10 Radiation pattern of the antenna in the H-plane. The points are interpolated with a 10th-order polynomial.

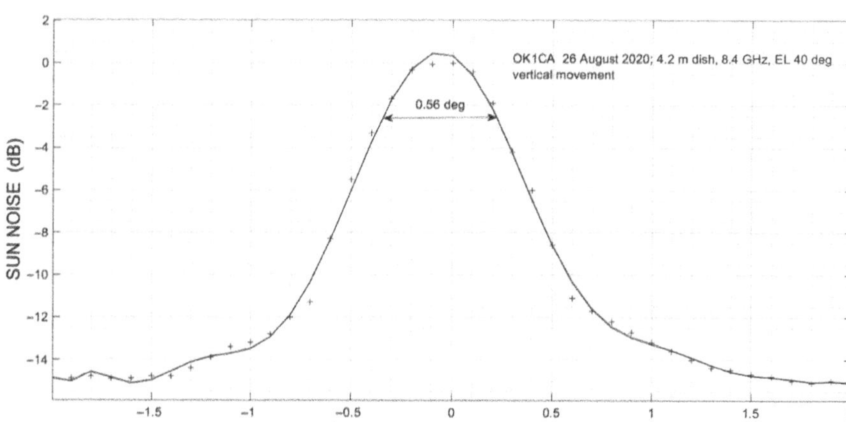

Figure 6.11 Radiation pattern of the antenna in E-plane. The points are interpolated with a 10th-order polynomial.

and the angular width of the Sun at 8.4 GHz

$$\Theta_S = 0.533 \cdot 10^{\frac{1}{\alpha_2}} = 0.533 \cdot 10^{\frac{1}{52.5}} = 0.557°,$$

$$K_2 = \frac{\ln(2) \cdot \left(\frac{0.557}{0.61}\right)^2}{1 - \exp\left[-\ln(2) \cdot \left(\frac{0.557}{0.61}\right)^2\right]} = 1.316.$$

For $Y_{\text{dB}} = 15\,\text{dB}$: $Y = 31.6$,

$$\frac{G}{T_s} = \frac{8 \cdot 3.14 \cdot 1.38 \cdot 10^{-23}(31.6 - 1)}{0.036^2 \cdot 200 \cdot 10^{-22}} \cdot 1 \cdot 1.316 = 539.1\,\text{K}^{-1},$$

$$\left(\frac{G}{T_s}\right)_{\text{dB}} = 10 \log 539.1 = 27.3\,\text{dB}\,\text{K}^{-1},$$

$$G_{\max\ \text{dB}} = 10 \log\left(\frac{35\,000}{\Theta^2_{3\,\text{dB}}}\right) = 10 \log\left(\frac{35\,000}{0.61^2}\right) = 49.7\,\text{dBi},$$

$$T_s = 10^{\frac{49.7-27.3}{10}} = 173.8\,\text{K},$$

$$\eta = \left(\frac{0.036}{3.14 \cdot 4.2}\right)^2 10^{\frac{49.7}{10}} = 0.69.$$

We can conclude that the system is very good with not much room for improvement. Maybe, T_s might be slightly lowered.

6.4 Measuring *G/T* Using Other Space Objects

As already mentioned, there are many other objects in the universe that significantly shine in the radio frequency spectrum, e.g. Cassiopeia A, Taurus A, Orion, Omega [11]. The radiation power densities of these sources are given in Table 6.2 [7]. Compared to the Sun and the Moon, those objects are much smaller, and can be used for direct measurements of G/T at larger apertures only. Comparison of Taurus A 476.0 Jy and the Moon $4.28 \times 10^{-22}\,\text{Wm}^{-2}\,\text{Hz}^{-1} = 42\,800\,\text{Jy}$ (see above) at a frequency of 10 368 MHz shows that Taurus is about 90x weaker.

$$K_{\text{Cdecy}} = 1 - \frac{0.97 - 0.3 \log f}{100} \tau(-;\,\text{GHz, year}) \tag{6.25}$$

Table 6.2 Radiation power densities of several other space sources.

Source	Radiation power density at frequency *f* [MHz]		*f* = 10 368 MHz
Cassiopeia A	$N_{\text{Cas}} = 10^{[5.745-0.770 \log f]} \cdot 10^{-26}$	(6.21)	449.7 Jy[a)]
Taurus A	$N_{\text{Tau}} = 10^{[3.794-0.278 \log f]} \cdot 10^{-26}$	(6.22)	476.0 Jy
Orion	$N_{\text{Ori}} = 10^{[3.317-0.204 \log f]} \cdot 10^{-26}$	(6.23)	314.6 Jy
Omega	$N_{\text{Ome}} = 1^{[4.056-0.378 \log f]} \cdot 10^{-26}$	(6.24)	345.2 Jy

a) Cassiopeia A is weakening as a radio source, and the value N_{Cas} from the table has to be therefore multiplied by the correction factor K_{Cdecy} [7].

Source: 7. ŠEKULJICA, D. (2017). Quality factor G/T direct measurement method of 7m–13m parabolic reflector antenna system with Moon as an RF source. Master thesis. University of Ljubljana.

where τ is the time in years since 1 January 1980. For 30 June 2020, it is therefore $\tau = 40.5$, $K_{\text{Cdecy}} = 0.73$ and thus $N^{\text{k}}_{\text{Cas}} = 449.7 \cdot 0.73 = 328.5$ Jy.

Other sources from Table 6.2 do not require such a correction.

References

1 Šekuljica, D., Badessi, S., Ferrante, M. et al. (2017). Considerations about the use of the Moon in X-band antenna G/T measurements. *Journal of Microelectronics, Electronic Components and Materials* 47(2): 79–89.

2 Daywitt, D.C. (1984). *An Error Analysis for the Use of Presently Available Lunar Radio Flux Data in Broadbeam Antenna-System Measurements.* NBS Technical Note 1073. National Bureau of Standards.

3 Daywitt, D.C. (1986). *10–60 GHz G/T Measurements Using the Sun as a Source – A Preliminary Study.* National Bureau of Standards.

4 Wait, D.F. and Daywitt, D.C. (1983). *Preliminary Examination of 20 GHz G/T Measurements of Earth Terminals.* National Bureau of Standards.

5 Welch, J. and DeBoer, D. (2004). *Expected Properties of the ATA Antennas.* ATA memo #66. https://www.seti.org/ata-memo-series (accessed 7 August 2024).

6 Wait, D.F. (1983). Precision measurement of antenna system noise using radio stars. *IEEE Transactions on Instrumentation and Measurement* 32(1): 110–116. https://doi.org/10.1109/TIM.1983.4315019.

7 Šekuljica, D. (2017). *Quality factor G/T direct measurement method of 7m–13m parabolic reflector antenna system with Moon as an RF source.* Master thesis. University of Ljubljana.

8 Solovey, A. and Mittra, R. (2008). Extended source size correction factor in antenna gain measurements. *Proceedings of the 38th European Conference on Microwaves,* Amsterdam: Netherlands (27–31 October 2008): 983–986. https://doi.org/10.1109/EUMC.2008.4751620.

9 RF Space: SpectraVueTM Software. http://www.rfspace.com/RFSPACE/SpectraVue.html (accessed 7 August 2024).

10 Tapping, K.F. (2013). The 10.7cm solar radio flux ($F_{10.7}$). *Space Weather* 11(7): 394–406. https://doi.org/10.1002/swe.20064.

11 Guidice, D.A. and Castelli, J.P. (1971). The use of extraterrestrial radio sources in the measurement of antenna parameters. *IEEE Transactions on Aerospace and Electronic Systems* 2: 226–234. https://doi.org/10.1109/TAES.1971.310353.

7

Parabolic Antennas

In microwave bands, weak signals are usually received by reflector antennas. The role of the reflector is played by a well-conducting surface in shape of a rotating paraboloid or its part.[1] If a transmitter is part of the system, the same antenna is usually used both for reception and for transmission. Such an antenna concentrates the plane wave power received by the aperture of the reflector to a single point (theoretically) called the focal point. The reflector significantly affects the quality of the antenna. Not only the size but also the accuracy and stability of the parabolic shape are important. In terms of reflectivity, fully metallic reflectors are the best ones. Due to the high wind resistance, mesh is often used instead of solid material in case of larger antennas, especially. The size of mesh elements has to be smaller than one tenth of the shortest wavelength at least. The same condition has to be kept in accuracy of the parabolic shape. Using a mesh reflector, a part of energy leaks behind the reflector which worsens the front-to-back radiation ratio and the equivalent antenna noise temperature (in case of receiving antennas). Even solid reflectors have problems. Larger microwave antennas are often heated for de-icing and the dilation of metallic materials. Cheap larger reflectors made by pressing aluminum can change shape over time (aluminum flows). Steel reflectors are pressed from thin sheets to reduce weight; their shape might be unstable as well. Laminate antennas with a pressed very fine mesh (with 1.0–1.5 mm mesh) are of relatively high quality in terms of accuracy and shape stability. Their quality is acceptable not only in the X-band, but in the Ka band as well.

7.1 Reflector Geometry and Its Irradiation

A fundamental arrangement of reflector antennas with a parabolic reflector is shown in Figure 7.1a,b. Each configuration has advantages and disadvantages.

1 In general, the shape of the reflector can be not only parabolic but also planar, angular, spherical...

Reception of Weak Radio Signals from Space, First Edition. Miroslav Kasal.

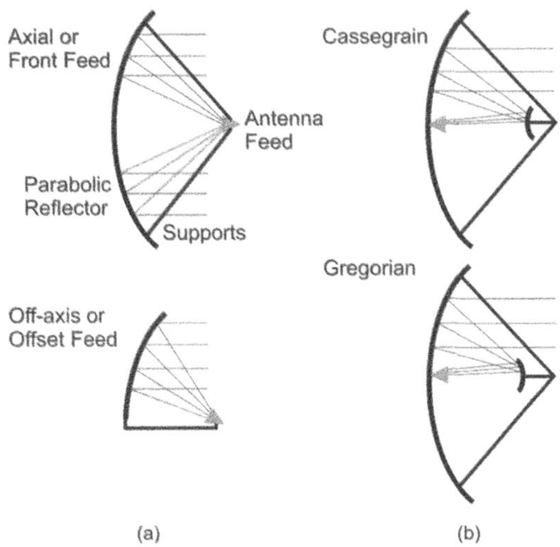

Figure 7.1 (a) Axisymmetric and offset arrangement; (b) dual-optic arrangement. *Source:* [1] /*Wikipedia/CC BY SA 4.0.*

7.1.1 Antenna with Rotationally Symmetric Reflector

An antenna with an axisymmetric reflector and a feed at the focal point, which is located on the axis of symmetry, is the simplest geometry (see Fig. 7.2). The ratio between the focal distance f and the aperture diameter d is the basic parameter of the reflector. A smaller f/d ratio is typical for deep reflectors with a large curvature and a larger f/d ratio is typical for shallow reflectors.

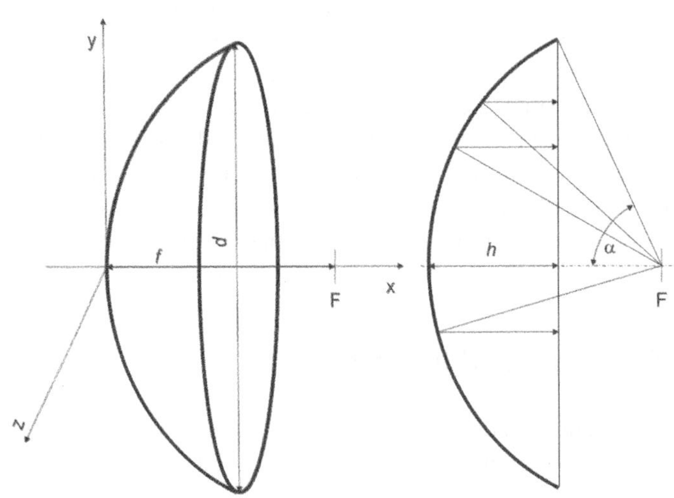

Figure 7.2 Geometry of the reflector in form of axisymmetric rotating paraboloid.

A parabolic surface in rectangular coordinates is described by the equation

$$y^2 + z^2 = 4fx \tag{7.1}$$

For $z = 0$ we get the equation of the parabola in the xy plane

$$y^2 = 4fx \tag{7.2}$$

when substituting $\left(\frac{d}{2}\right)^2 = 4fh$, from there then

$$f = \frac{d^2}{16h} \tag{7.3}$$

7.1.2 Antenna with Offset Reflector

The offset reflector antenna geometry is shown in Figure 7.3. The reflector is formed by a part of the surface of a larger parabolic surface. In xy plane, the surface is given by Eq. (7.2). The surface is irradiated at an offset angle θ.

The figure shows the main advantage of offset arrangement. The feed is not irradiated by the incident wave and does not shield the reflector as in case of axially symmetric antennas. At low elevation angles of the antenna,[2] the feed is directed upwards and the effect of spillover on noise temperature is small (this space is mostly "cold"). The offset reflector is usually oval. At the offset angle, aperture of an elliptically shaped reflector appears circular (see Fig. 7.4). For the same reason, the shorter axis of the ellipse d is always given as the aperture diameter of the offset antenna. For the depth of the offset reflector h and the major axis of the ellipse

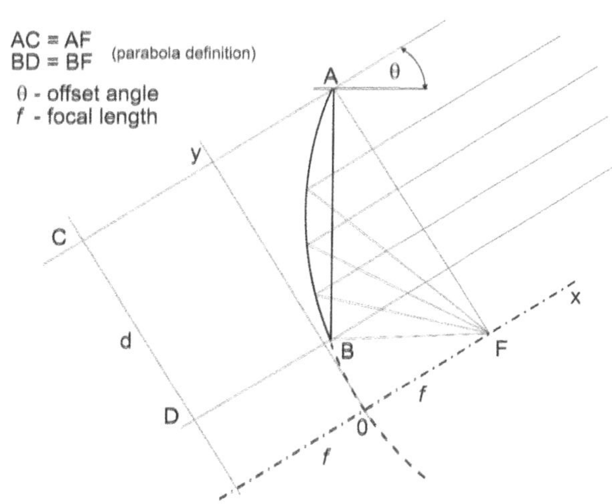

Figure 7.3 Geometry of a parabolic antenna with an offset reflector.

2 Sensitivity of the receiving system at low elevation is usually the most critical.

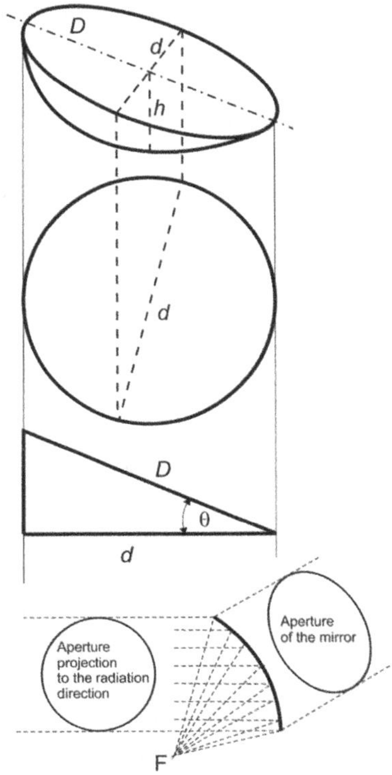

Figure 7.4 Projection of the oval aperture to the direction of radiation [2].

D, the offset angle θ in deg (°) is [2] (see Fig. 7.4):

$$\theta = \arccos\left(\frac{d}{D}\right) \tag{7.4}$$

and for focal distance f in m [2]

$$f = \frac{d^3}{16hD} \tag{7.5}$$

On the contrary, a more difficult focusing is the disadvantage of offset reflectors. With an axisymmetric antenna, a focal point is sought by moving the feed along the antenna axis (a single degree of freedom) [3]. With an offset antenna, there are at least three degrees of freedom. For feed position, antenna manufacturers therefore specify four measures A, B, C, D (see Fig. 7.5) [4]. There are two marks on the reflector the measures are related to.

For the irradiation of offset reflectors, the same principles apply as for axisymmetric ones.[3] Since offset antennas are manufactured with a larger f/d ratio, feeds should have narrower radiation characteristics.

3 For the feed, the aperture appears circular under the offset angle.

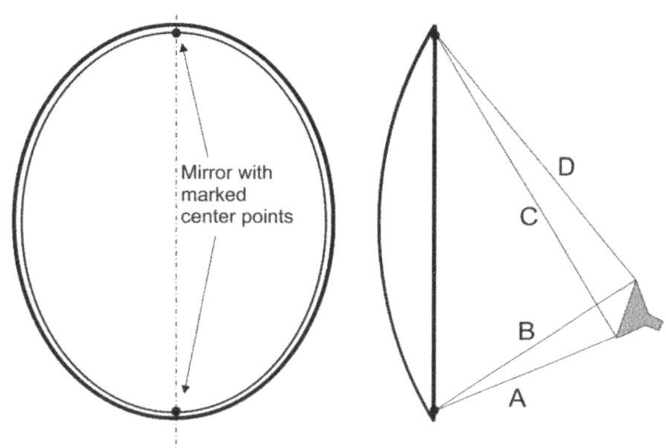

Figure 7.5 Measures to focus offset antenna. *Source: 4. PRODELIN. Technical Bulletin. Number 97-01x, 1 October 1997.*

7.1.3 Antennas with Dual Optics

Larger antenna systems are often constructed with "dual optics" (see Figs. 7.6a,b). The systems are supplemented with a secondary reflector, which helps to achieve nearly uniform irradiation of the primary reflector. In Cassegrain arrangement, the secondary reflector is a convex surface (a hyperboloid of revolution, usually). In Gregorian arrangement, the secondary reflector is a concave surface (an ellipsoid, usually). Both solutions allow to "hide" the feed itself – either behind the primary reflector (see Fig. 7.6a) or even further away from the radiation illuminating the aperture (see Fig. 7.6b). However, the demands on the accuracy and surface quality of secondary reflectors are considerable. BWG (Beam Wave Guide) antennas are the special case. BWGs are used in DSN (Deep Space Network) stations. The signal from the focal point of the primary reflector of BWG is guided by a system of reflectors to the point of signal detection (see Chapter 13).

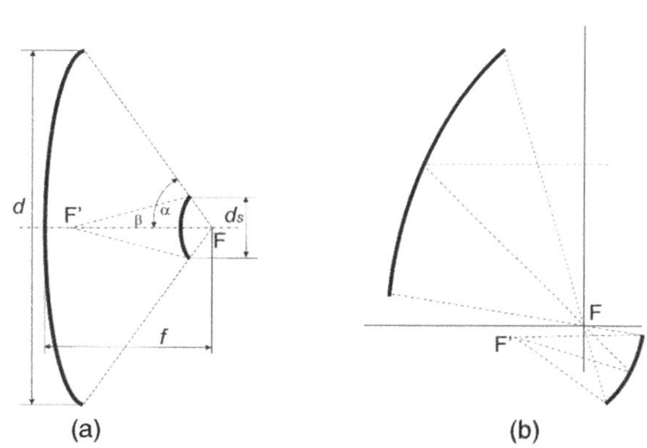

Figure 7.6 (a) Cassegrain configuration of axisymmetric antenna; (b) Gregorian configuration of offset antenna.

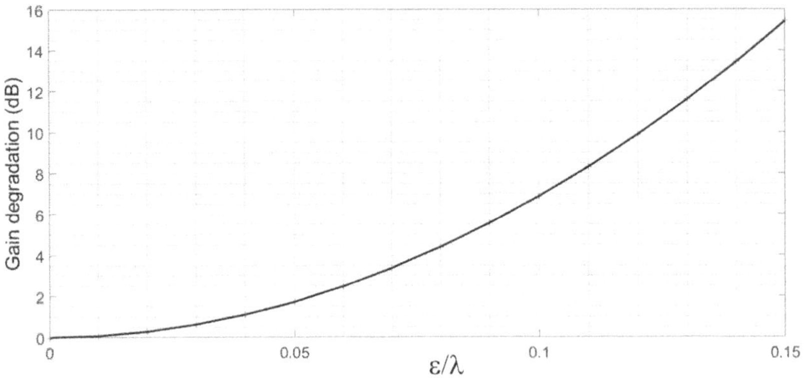

Figure 7.7 Parabolic antenna gain degradation depending on the effective value of deviations ε/λ from an ideal paraboloid.

7.1.4 Accuracy of Parabolic Reflectors

As already mentioned, accuracy of a parabolic reflector affects the overall efficiency of the antenna. In 1952, John Ruze published Eq. (7.6) [5] which quantifies the degradation of parabolic antenna gain (2.16) as a function of the root-mean-square (RMS) deviations ε of a real reflector from an ideal paraboloid

$$G_{\text{eff}} = G_{\text{max}}\, \mathrm{e}^{-\left(\frac{4\pi\varepsilon}{\lambda}\right)^2} \tag{7.6}$$

and in a logarithmic scale

$$G_{\text{eff dB}} = G_{\text{max dB}} - 685.81\left(\frac{\varepsilon}{\lambda}\right)^2 \tag{7.7}$$

The chart in Figure 7.7 shows how fatal the consequences of an inaccurate parabolic reflector are. At the effective deviation $\varepsilon = 0.1\lambda$, the antenna gain degradation is higher than 6 dB.

7.2 Feeds

For half the aperture angle from the focus α (see Fig. 7.2)
$\alpha = \arctan\left(\frac{d/2}{f-h}\right)$ and after substituting from (7.3)

$$\alpha = \arctan\left[\frac{8(f/d)}{16(f/d)^2 - 1}\right] \tag{7.8}$$

Graphically, the dependence is shown in Figure 7.8. Antennas with the same f/d ratio have the same angle 2α, and the same feed can be used.

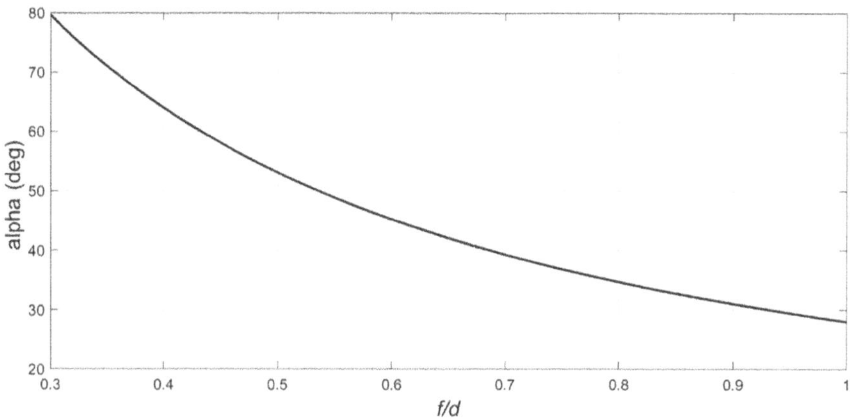

Figure 7.8 The half angle of the aperture measured from the focal point, depending on the f/d ratio.

Next to the quality of the reflector, its optimum irradiation is very important as well.[4] When handling with weak signals (equivalent noise temperature T_S is the critical parameter), aperture irradiation with 10 dB decrease at the edges is usually selected (edge taper). Let us recall the relation (2.17) with note[7] in Chapter 2. In general, we can write

$$\theta_{3\text{dB}} = k\frac{\lambda}{d} \quad (^\circ; \text{m}) \tag{7.9}$$

In Chapter 2, we considered $k = 70$ [6] Gaussian approximation of the beam of the irradiator was assumed. In [7], an expression for the proportionality constant k is given

$$k = 58.96(1 + 0.0107E_t) \quad (^\circ; \text{dB}) \tag{7.10}$$

Here, E_t is the decrease of power at the edge of the aperture in dB.

Graphical representation of Eq. (7.10) in Figure 7.9 shows that 10 dB power decrease at aperture edges corresponds to the value $k = 65$. This value is also reported by some sources, e.g. [8]. On the contrary, the value $k = 70$ corresponds to 17 dB decrease, approximately. If the value of $E_t = 10$ dB is taken as realistically optimal, then a slightly higher gain can be achieved at lower values of E_t. A higher equivalent noise temperature (due to spillover) is the price we pay for. On the opposite, a higher E_t decreases the noise temperature at the expense of gain. Irradiation of a large reflector around its center only is a typical example. Another important condition requires having the phase center of the feed in the focal point of the reflector exactly [9]. If a horn is the base of the feed, the phase center is around the top of the horn [10, 11].

4 Uniform wave without spillover irradiates the reflector ideally. Coefficient $k = 59$ in Eq. (7.9) corresponds to that.

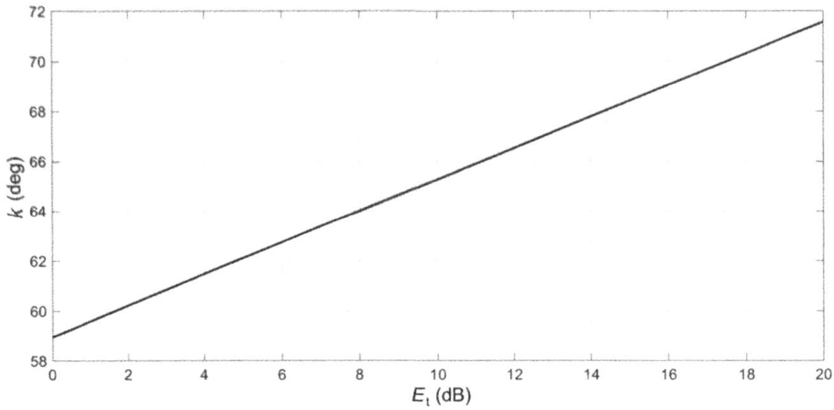

Figure 7.9 Graphic representation of Eq. (7.10): Dependence of the proportionality constant k on the decrease of power at the edge of the aperture E_t.

Selected examples of 10 GHz feeds and 24 GHz ones follow [12–14]. Different f/d ratios correspond to real, used reflectors. Due to reasons discussed in Chapter 11, feeds are designed for linear polarization.

7.2.1 Open Circular Waveguide in 10 GHz Band

Radiation characteristics of an open circular waveguide [12] were simulated for interest. For the dominant TE_{11} mode and frequency 10.368 MHz, a circular waveguide can be made from a copper tube with an internal diameter 20 mm (0.692 λ). The outer diameter of the waveguide is 22 mm. Simulation results are shown in Figures 7.10 and 7.11.

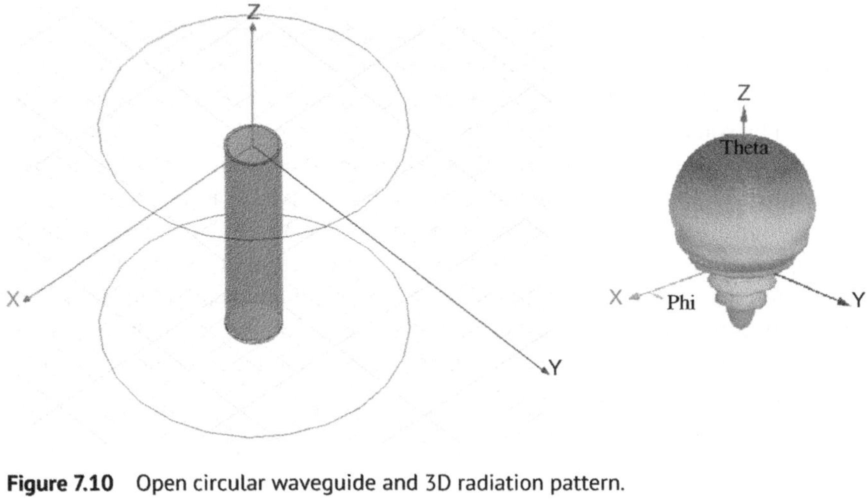

Figure 7.10 Open circular waveguide and 3D radiation pattern.

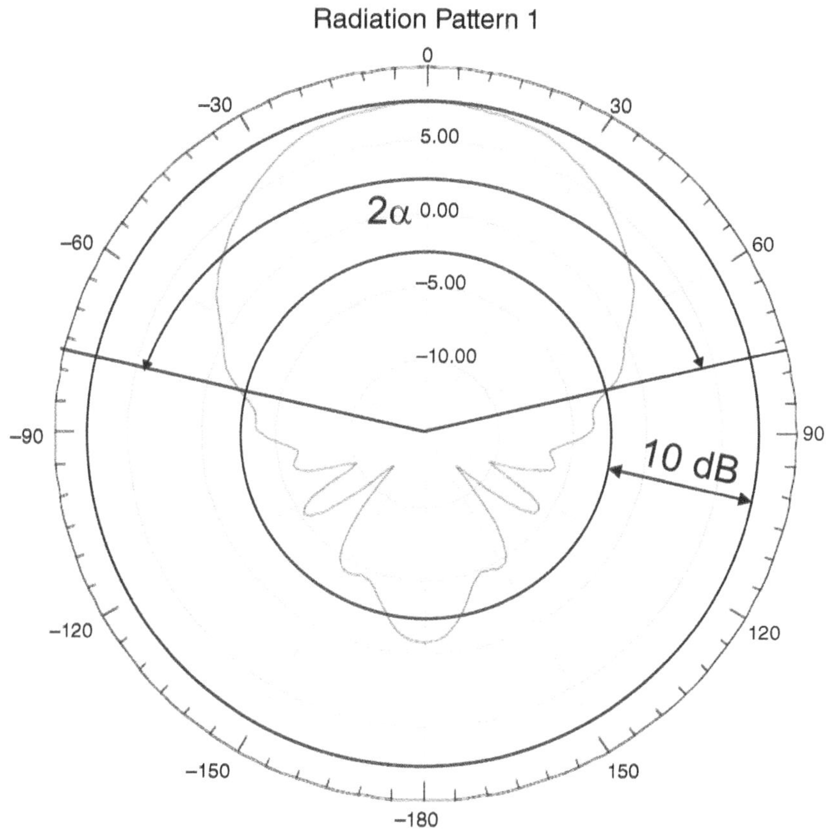

Figure 7.11 2D radiation pattern of open waveguide and 2α angle for 10 dB decrease.

Since radiation characteristics are rotationally symmetric, the angle for aperture irradiation 2α can be determined from radiation pattern in the xz plane (see Fig. 7.11). For 10 dB decrease at the edges of the aperture, $\alpha = 82°$. By an open waveguide, a deep reflector with $f/d < 0.3$ could be irradiated (see Fig. 7.8).

7.2.2 Open Circular Waveguide with Simple Choke

An open waveguide is usually supplemented with a simple choke, see Fig. 7.12 [12]. The choke of suitable dimensions resists the currents otherwise flowing along the surface of the waveguide. Moving gently, the front of radiation characteristics can be primarily shaped (see Fig. 7.13).

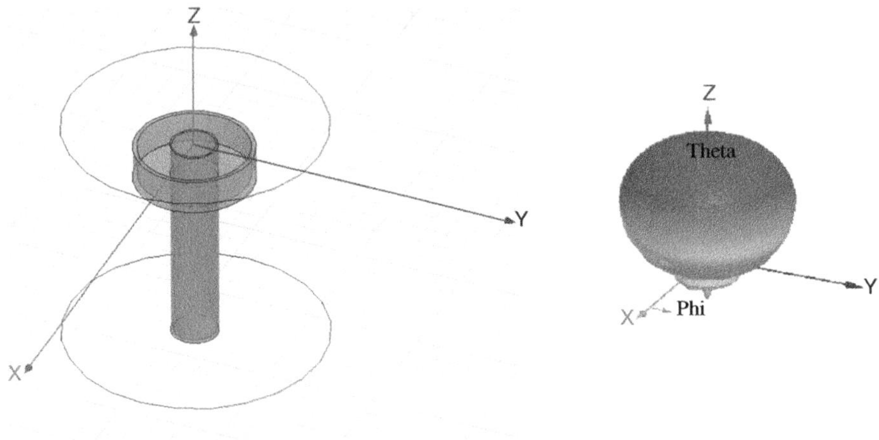

Figure 7.12 Waveguide with choke: The inner radius of the collar is 25.5 mm and the inner height is 14.5 mm. The shift of the inner wall of the choke relative to the face of the waveguide is 19 mm.

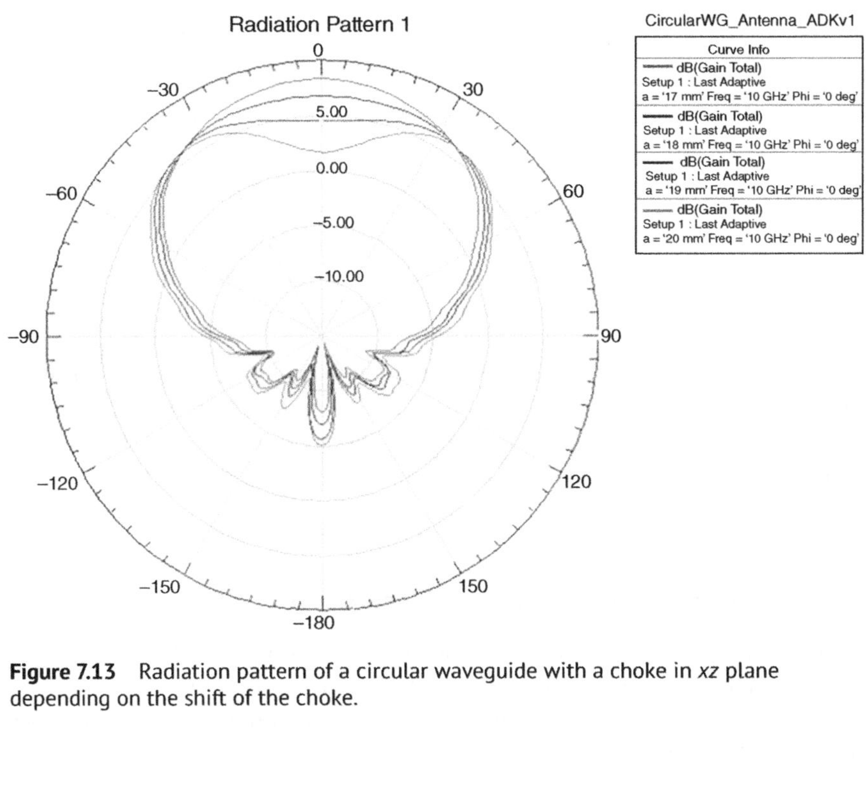

Figure 7.13 Radiation pattern of a circular waveguide with a choke in *xz* plane depending on the shift of the choke.

7.2.3 Scalar Feed

This feed type [12] is often used to irradiate reflectors of $f/d \approx 0.4$–0.5, see Figs. 7.14–7.16.

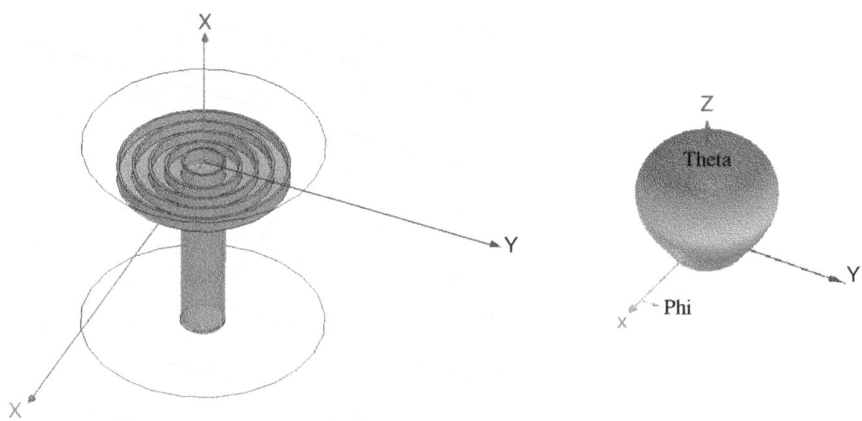

Figure 7.14 Scalar feed and 3D radiation diagram.

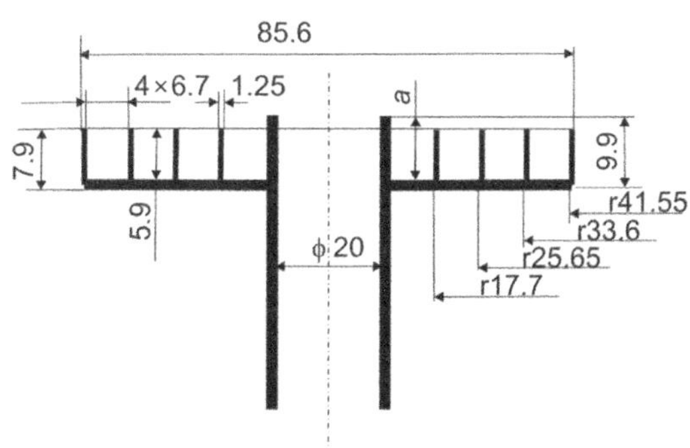

Figure 7.15 Schematic drawing of the simulated scalar feed at 10 GHz.

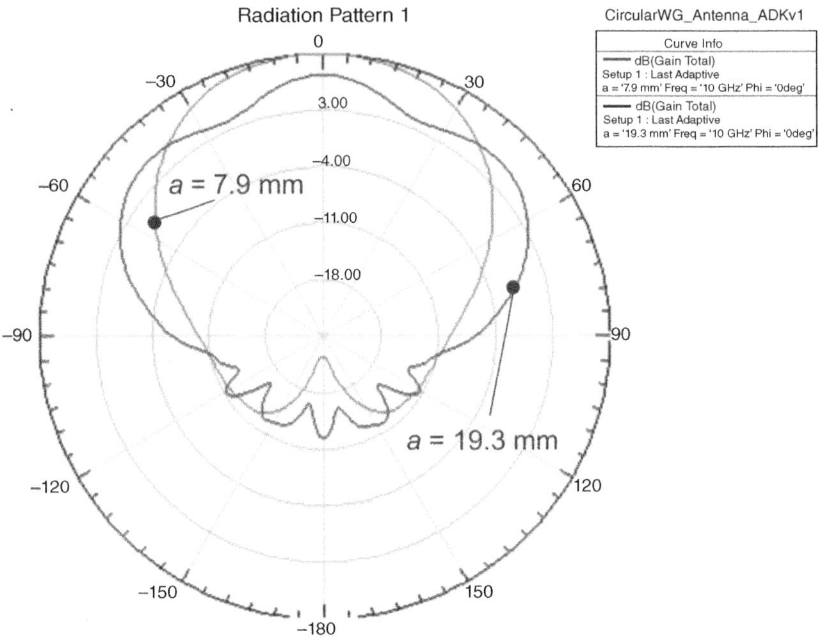

Figure 7.16 Radiation diagram of a scalar feed in a *xz* plane for two values of the shift *a* of the ring relative to the face of the circular waveguide.

7.2.4 Horn with Rings

The horn with rings designed in [12] was analyzed as well. This feed might be suitable for reflectors of $f/d = 0.6$, see Figs. 7.17–7.18.

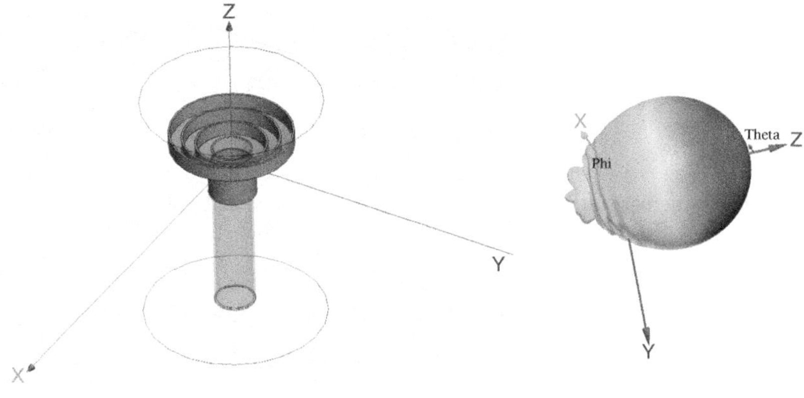

Figure 7.17 Model and radiation pattern of the horn feed with rings.

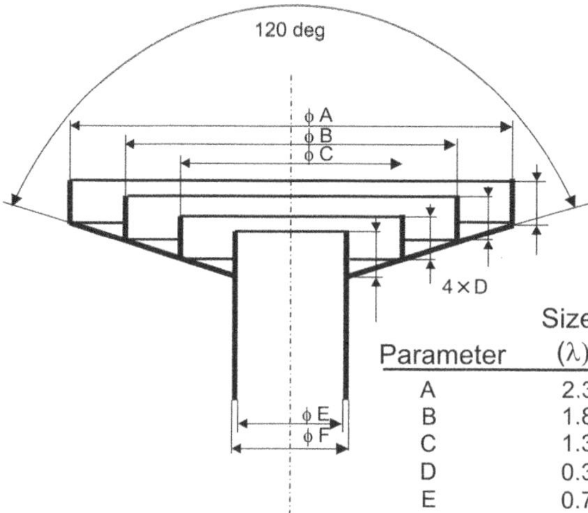

Parameter	Size (λ)	Size for 10.4 GHz (mm)
A	2.3	66.5
B	1.8	52.0
C	1.3	37.5
D	0.3	8.7
E	0.7	20.0
F	0.9	26.0

Figure 7.18 Dimensions of the horn feed with rings (wall thickness 1 mm).

7.2.5 Horn for 10 GHz and Offset Reflector with $f/d = 0.8$

A shallow offset reflector with $f/d = 0.8$ has an aperture angle from the focal point $2\alpha \approx 70^{\circ}$. A horn with rings (corrugated horn) appears to be a suitable feed, but with somewhat larger dimensions [15]. Dimensions of Prodelin's original Ku-band feed were adapted to allow shifts along a circular waveguide of 20/22 mm diameter. Results at 10.4 GHz are very good (see Figs. 7.19–7.21).

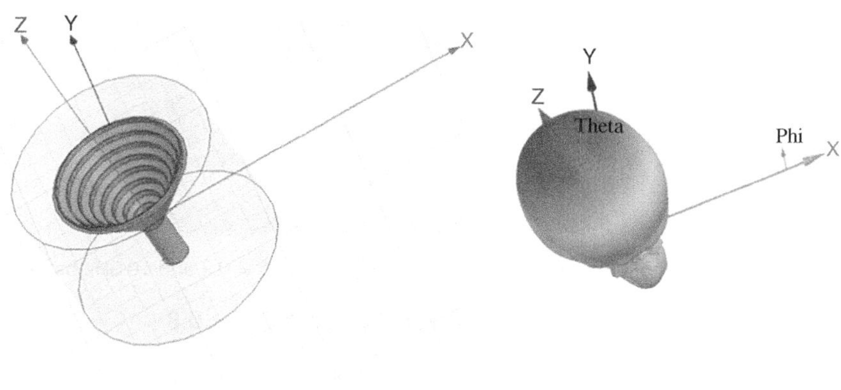

Figure 7.19 Feed of offset reflector $f/d = 0.8$ for 10.4 GHz and 3D radiation pattern.

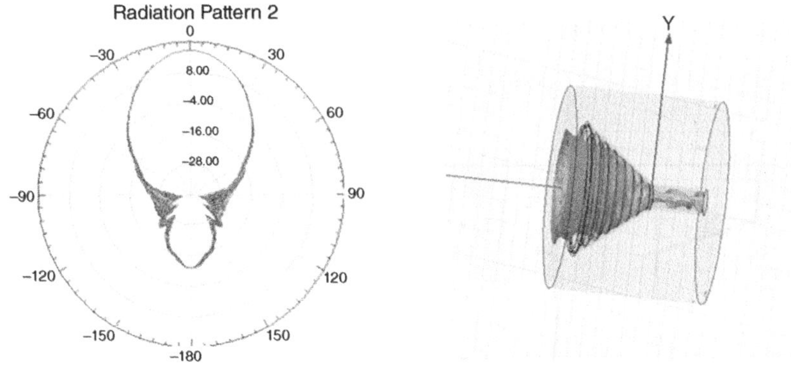

Figure 7.20 2D radiation diagram in different planes and E-field inside the feed. The phase center is on the axis z 7 mm inside the cylindrical waveguide.

Figure 7.21 Modified Prodelin's feed for offset reflector $f/d = 0.8$ and 10 GHz band.

7.2.6 Horn for 24 GHz and Offset Reflector with $f/d = 0.8$

The horn for 24 GHz and offset reflector with $f/d = 0.8$ are designed [13], fabricated and tested (see Figs. 7.22 and 7.23). Measured results agree well with the simulated ones (see Figs. 7.24 and 7.25).

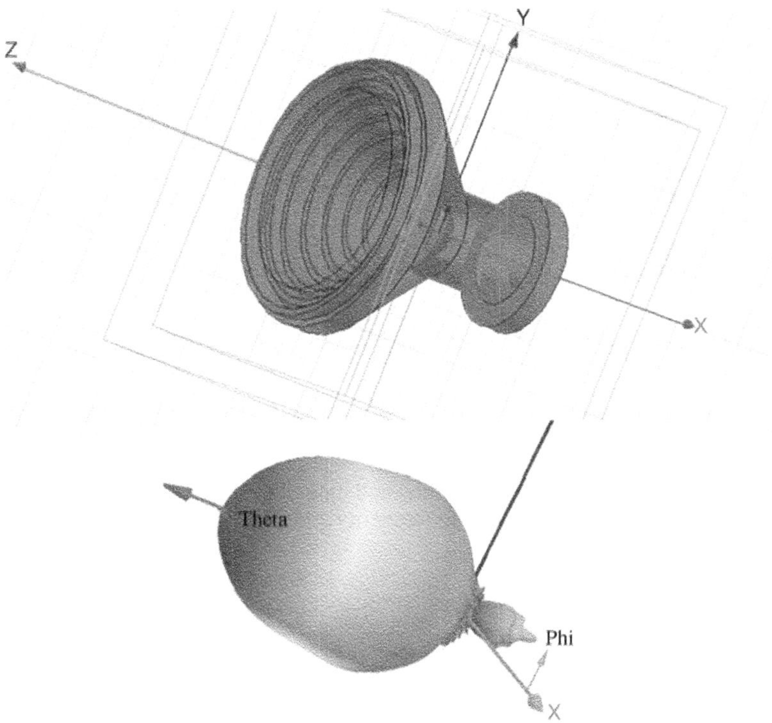

Figure 7.22 Feed of corrugated horn type for offset reflector with $f/d = 0.8$ at 24 GHz.

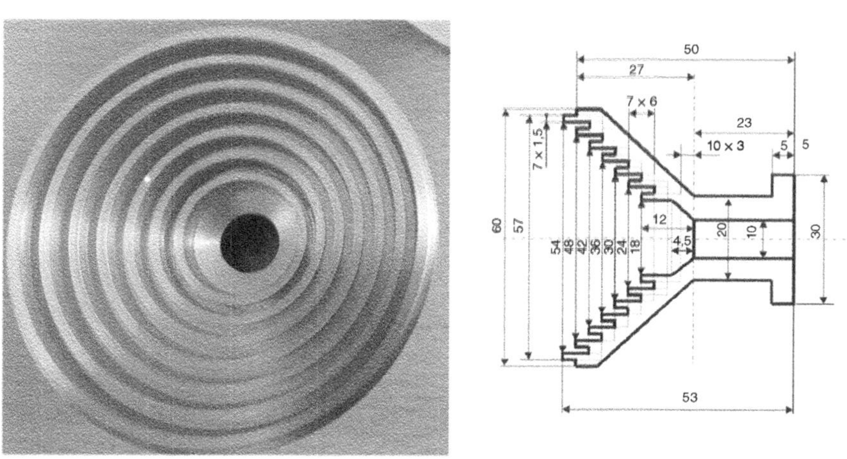

Figure 7.23 Feed implementation and cut with dimensions.

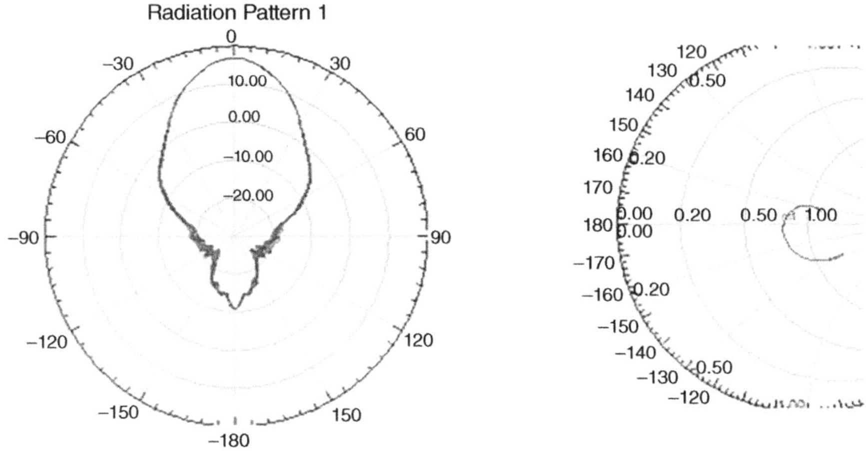

Figure 7.24 Radiation characteristics Gain-total and input impedance of feed normalized to 50 Ω. Point m1 corresponds to frequency 24.048 MHz.

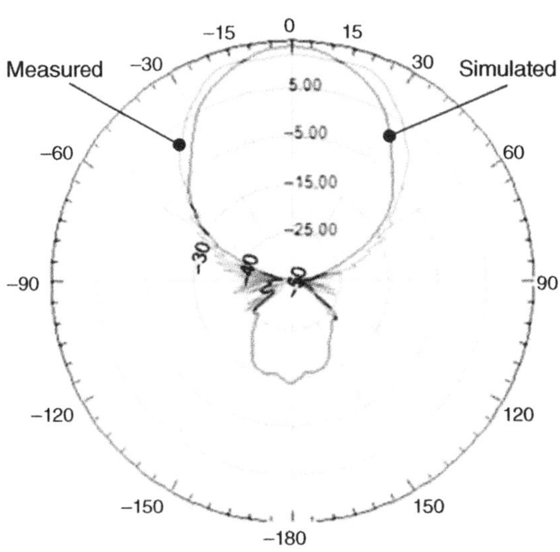

Figure 7.25 Simulated radiation characteristics Gain-L3x[5] and the corresponding characteristics measured in an anechoic chamber.

7.2.7 Horn for 24 GHz and Offset Reflector with $f/d = 0.6$

The horn for 24 GHz and offset reflector with $f/d = 0.6$ are designed [14], fabricated and tested (see Figs. 7.26 and 7.27). Measured results agree well with the simulated ones (see Figs. 7.28 and 7.29).

5 Radiation characteristics Gain-L3x considers cross-polarization, corresponding to measurements in an anechoic chamber.

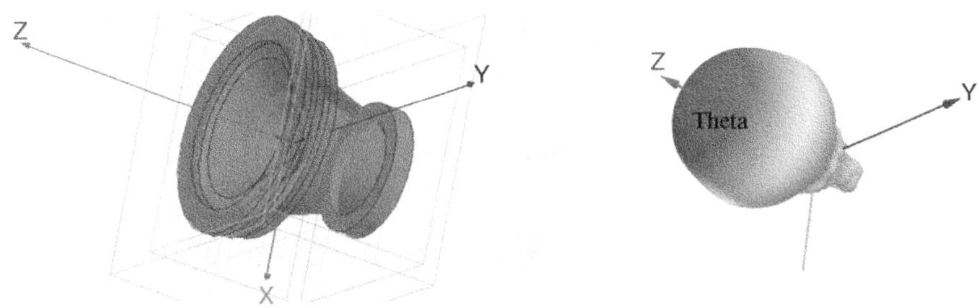

Figure 7.26 Feed of corrugated horn type for offset reflector with $f/d = 0.6$ at 24 GHz.

Figure 7.27 Feed implementation for the reflector with $f/d = 0.6$ and drawing with dimensions. *Source (photo): OK 1 DFC.*

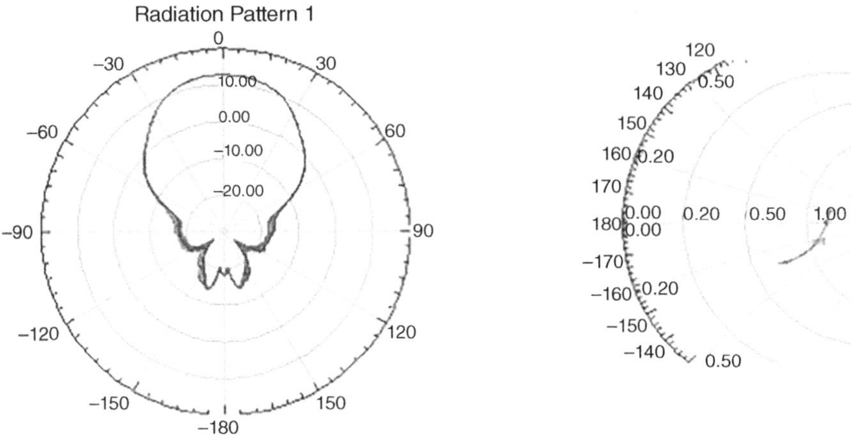

Figure 7.28 Radiation characteristics Gain-total and input impedance of feed normalized to 50 Ω. Point m1 corresponds to frequency 24.048 MHz.

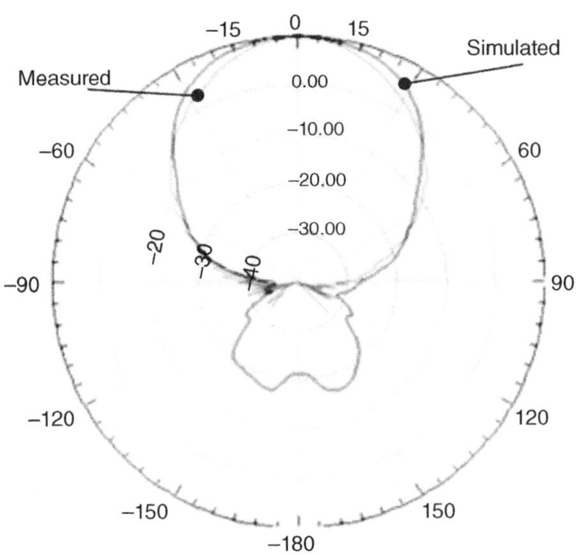

Figure 7.29 Simulated radiation characteristics GainL3x and corresponding characteristics measured in an anechoic chamber.

7.2.8 Standard Horn for 1.3 GHz and Offset Reflector with $f/d = 0.8$

In 1.3 GHz band, circular polarization is often used, and a feed with septum [16] is a suitable solution. A typical solution has radiation characteristics suitable for a reflector with $f/d \approx 0.4$–0.5. A new solution was therefore sought to modify the feed for a reflector with $f/d = 0.8$. The resultant horn, its implementation and achieved radiation patterns can be seen in Figures 7.30–7.33 [17].

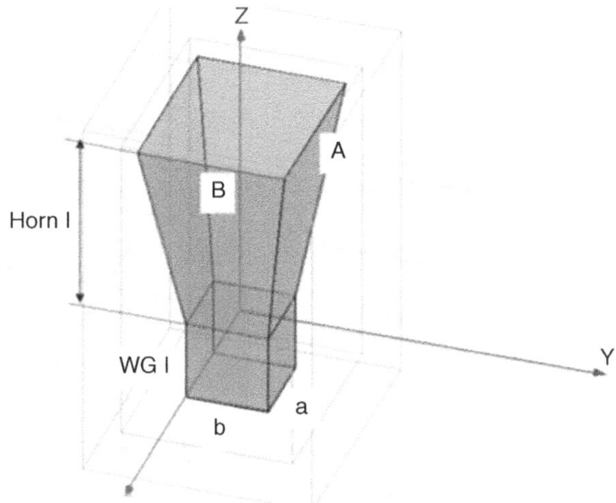

Figure 7.30 Dimensions of the 1.3 GHz horn extending the standard feed with septum [17]: a = b = 150 mm, WGI = 150 mm, A = 335 mm, B = 270 mm, Horn I = 400 mm.

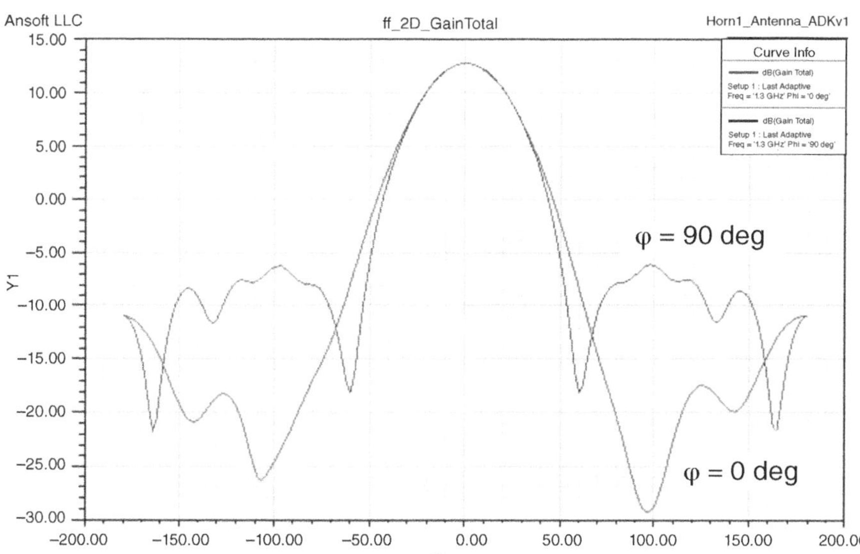

Figure 7.31 Radiation characteristics of the feed with septum extended by the above-described horn.

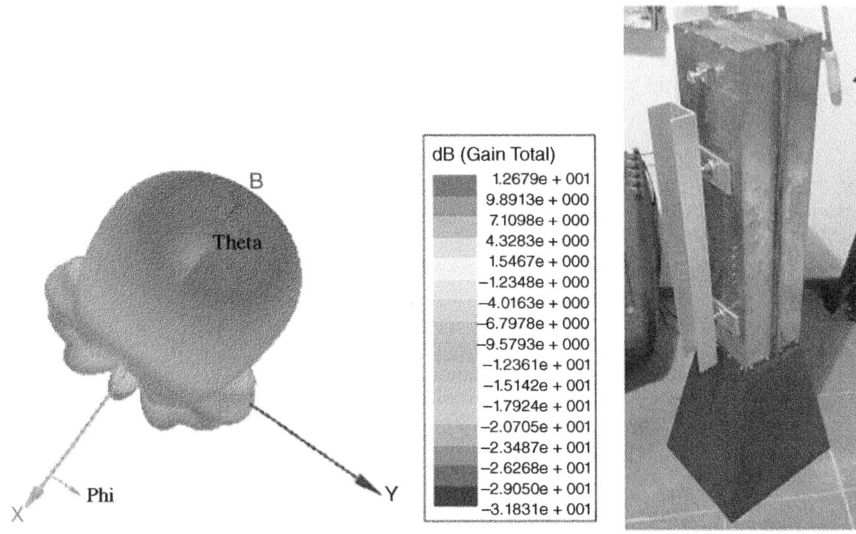

Figure 7.32 3D radiation characteristics of the feed with septum extended by the above-described horn and the feed implementation.

Figure 7.33 The system was tested with 1.8 m offset reflector. By measuring SN/CS = 7 dB at SFU = 70, very good properties of the antenna were verified in the 1.3 GHz band, with respect to computations. *Source (photo): OK 1 DFC.*

References

1 Wikipedia. *Parabolic Antenna*. https://en.wikipedia.org/wiki/Parabolic_antenna (accessed 7 August 2024).

2 Legon, J.A.R. *Calculation of the Focal Length of an Offset Satellite Dish Antenna*. https://www.academia.edu/32059393/Calculation_of_the_Focal_Length_of_an_Offset_Satellite_Dish_Antenna (accessed 7 August 2024).

3 Balanis, C.A. (2005). *Antenna Theory: Analysis and Design*. 3rd ed. Hoboken, NJ: Wiley.

4 Prodelin. (1997). Technical Bulletin. Number 97-01x, 1 October.

5 Wikipedia. *Ruze's Equation*. https://en.wikipedia.org/wiki/Ruze%27s_equation#cite_note-1 (accessed 7 August 2024).

6 Nelson, R.A. (1999). Antennas: The interface with space. *Via Satellite*. https://www.scribd.com/document/239670143/Satellite-Antennas (accessed 7 August 2024).

7 Šekuljica, D. (2017). Quality factor G/T direct measurement method of 7m–13m parabolic reflector antenna system with Moon as an RF source. Master thesis. University of Ljubljana.

8 International Telecommunication Union (2002). *Handbook on Satellite Communications*. 3rd ed. Wiley. ISBN 978-0-471-22189-0.

9 Hazdra, P. and Mazánek, M. (2015). *Reflector Antennas and Their Feeds*. https://http://elmag.fel.cvut.cz/sites/default/files/users/hazdrap/files/Reflector_antennas.pdf (accessed 7 August 2024).

10 Muehldorf, E.I. (1970). The phase center of horn antennas. *IEEE Transactions on Antennas and Propagation* 18(6): 753–760. https://doi.org/10.1109/TAP.1970.1139799.

11 Zechmeister, J. (2018). Antenna system for automotive radar (in Czech). Master thesis. Brno University of Technology. https://www.vut.cz/studenti/zav-prace/detail/110456 (accessed 7 August 2024).

12 Kasal, M. (2012). A study of parabolic mirror feeds in 10 GHz band (in Czech). http://www.urel.feec.vutbr.cz/esl/files/EME/Doc/Ozarovace_1.pdf (accessed 7 August 2024).

13 Kasal, M. (2012). 24 GHz conical horn for offset dish $f/d = 0.8$. http://www.urel.feec.vutbr.cz/esl/files/EME/Doc/Conical_Horn_0_8_OK2AQ.pdf (accessed 7 August 2024).

14 Kasal, M. (2012). 24 GHz feed for a dish $f/d = 0.6$. http://www.urel.feec.vutbr.cz/esl/files/EME/Doc/Conical_Horn_0_6_OK2AQ.pdf (accessed 7 August 2024).

15 Kasal, M. (2018). *Offset Dishes for Microwave EME*. https://www.radio.feec.vutbr.cz/esl/files/EME/Doc/Offset%20Dishes%20for%20MW%20EME.pdf (accessed 7 August 2024).

16 Samek, Z. (2022). Septum polarize – transformer, feed and all about. http://www.ok1dfc.com/eme/emeweb.htm (accessed 7 August 2024).

17 Kasal, M. (2012). Extended 23 cm septum feed for *f*/*d* = 0.8 dish. http://www.urel.feec.vutbr.cz/esl/files/EME/Doc/OK2AQ_23cm.pdf (accessed 7 August 2024).

8

High-Frequency Transmission Lines – Waveguides

Microwave signals are often transmitted through lines[1] with transverse dimensions significantly smaller than the wavelength. In the transverse plane, the electromagnetic field is stationary. The wave propagating along such a line is called transverse electric-magnetic (TEM) one. These lines include so-called two-wire lines – coaxial, microstrip, stripline, and others. In case of microstrip and stripline structures, the wave is called quasi-TEM. If the transverse dimensions of lines are comparable to the wavelength, the field distribution has a wave nature not only in the longitudinal direction but in the transverse one as well. Such a system is called a waveguide. Along the waveguide, only a wave whose frequency higher than a certain limiting – the so-called cutoff or critical frequency – can propagate. In a broader sense of the word, waveguides can be classified by the general term "transmission lines." In general, any metal tube can be the waveguide. For practical reasons, waveguides of rectangular or circular cross-section are used. Waveguides can be both homogeneous and inhomogeneous – both transversely and longitudinally [1, 2].

Rectangular and circular waveguides (see Fig. 8.1) are the basic shapes. Along an ideal hollow metal waveguide, waves can propagate without loss if their wavelength λ is shorter than the cutoff wavelength, λ_c (the frequency is higher than the critical frequency, f_c). Hence, the waveguide behaves like a high-pass filter with very steep characteristics. Waves propagate in the waveguide by reflections from walls. Therefore, the phase velocity, v_p, is higher than the speed of electromagnetic wave propagation in free space with permittivity ε and permeability μ.

The speed of electromagnetic wave propagation in free space is given by[2]

$$v = \frac{1}{\sqrt{\varepsilon\mu}}. \tag{8.1}$$

1 This chapter was dominantly adopted from [1].
2 For dry air, $v = c$, where c is the speed of light in vacuum (3×10^8 m s^{-1}).

Reception of Weak Radio Signals from Space, First Edition. Miroslav Kasal.

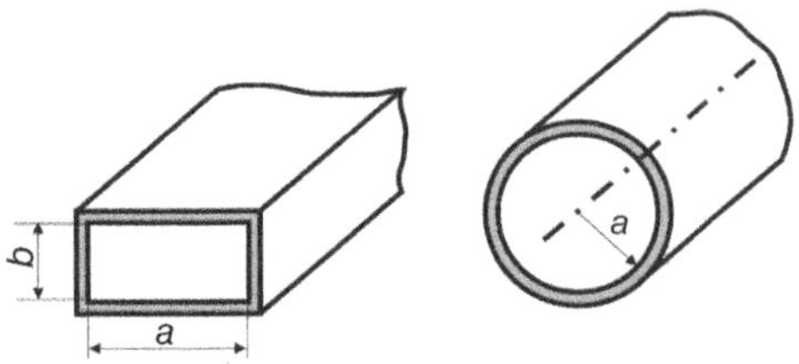

Figure 8.1 Rectangular and circular waveguide.

Points of the same phase propagate in the waveguide by the phase velocity

$$v_{\mathrm{p}} = \frac{v}{\sqrt{1-\left(\frac{f_{\mathrm{c}}}{f}\right)^{2}}} = \frac{v}{\sqrt{1-\left(\frac{\lambda}{\lambda_{\mathrm{c}}}\right)^{2}}}. \tag{8.2}$$

Points of the same phase of the modulation envelope propagate in the waveguide by the group velocity v_τ

$$v_\tau = v\sqrt{1-\left(\frac{f_{\mathrm{c}}}{f}\right)^{2}} = v\sqrt{1-\left(\frac{\lambda}{\lambda_{\mathrm{c}}}\right)^{2}}. \tag{8.3}$$

Wavelength in the waveguide is the distance the wave passes by phase velocity v_{p} in one period $T = 1/f$, i.e.

$$\lambda_{\mathrm{g}} = \frac{\lambda}{\sqrt{1-\left(\frac{\lambda}{\lambda_{\mathrm{c}}}\right)^{2}}}, \tag{8.4}$$

where $\lambda = v/f$. In the passband, the wavelength in the waveguide is always longer than the wavelength in vacuum. This is due to the higher phase velocity and longer path the wave takes when reflected from the walls of the waveguide.

The solution of Maxwell's equations shows that the total electromagnetic field inside the waveguide can be expressed as a superposition of transversely magnetic TM and transversely electric TE waves. In a waveguide, TE and TM waves can exist separately, can be excited separately, and can propagate independently. Each TM and TE wave represents a separate transmission channel of the waveguide.

The characteristic impedance of the waveguide is defined as the ratio of complex amplitudes of the transverse components of the electric and magnetic field intensities in the waveguide. In the passband, the characteristic impedance is real:

$$Z_{0\mathrm{TM}} = \sqrt{\frac{\mu}{\varepsilon}}\sqrt{1-\left(\frac{f_{\mathrm{c}}}{f}\right)^{2}} \text{ and } Z_{0\mathrm{TE}} = \frac{\sqrt{\frac{\mu}{\varepsilon}}}{\sqrt{1-\left(\frac{f_{\mathrm{c}}}{f}\right)^{2}}}. \tag{8.5}$$

In the passband, losses in the waveguide are caused by the finite conductivity of waveguide walls and by losses in dielectrics. Losses in waveguide walls are characterized by the penetration depth, δ, or the so-called high-frequency surface resistance, R_s:

$$\delta = \sqrt{\frac{2}{\omega \mu_w \sigma_w}} \quad \text{and} \quad R_s = \frac{1}{\delta\, \sigma_w} = \sqrt{\frac{\omega\, \mu_w}{2\, \sigma_w}}. \tag{8.6}$$

Here, μ_w and σ_w are the permeability and conductivity of waveguide wall materials.

Losses in dielectrics are characterized by nonzero leakage conductivity of the dielectrics σ inside the waveguide. The attenuation factor, α_d, in neper per meter (Np m^{-1}) is

$$\alpha_d = \frac{\omega\, \sqrt{\varepsilon \mu}}{2} \frac{\tan\delta}{\sqrt{1 - \left(\frac{f_c}{f}\right)^2}}. \tag{8.7}$$

Here, $\tan\delta = \sigma/(\omega\varepsilon)$ is the loss factor of the dielectric material. If the waveguide is filled with dry air, the losses are mostly negligible [3, 4].

8.1 Rectangular Waveguides

In a waveguide of rectangular cross-section (see Fig. 8.2), an infinite number of different TM waves and an infinite number of different TE waves can exist. Each mode is characterized by two nonnegative so-called mode numbers, m and n. Mode numbers can only have values that meet the boundary conditions on the conductive surface of the waveguide. This condition implies that TM_{mn} modes must have both nonzero numbers and TE_{mn} modes can have one zero number. For critical frequencies (for critical wavelengths) of the rectangular waveguide

$$f_c = \frac{1}{2\pi\, \sqrt{\varepsilon\mu}} \sqrt{\left(\frac{m\pi}{a}\right)^2 + \left(\frac{n\pi}{b}\right)^2} \quad \text{or} \quad \lambda_c = \frac{2\pi}{\sqrt{\left(\frac{m\pi}{a}\right)^2 + \left(\frac{n\pi}{b}\right)^2}}. \tag{8.8}$$

Figure 8.2 Rectangular waveguide.

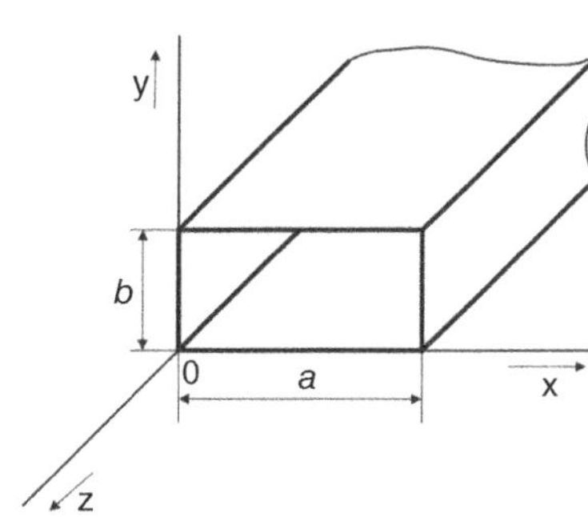

For given waveguide dimensions, different modes of both types of waves have different cutoff frequencies and different phase velocities, v_p (Eq. 8.2), different group velocities v_τ (Eq. 8.3), different wavelengths λ_g (Eq. 8.4), different characteristic impedances Z_{0TE}, Z_{0TM} (Eq. 8.5), and transmit different active power.

From all the modes, the so-called dominant mode is the most important one. The dominant mode has the lowest cutoff frequency f_c, or equivalently, the largest wavelength λ_c. Equation (8.8) shows that the dominant mode of the rectangular waveguide is TE_{10} (see Fig. 8.3) with cutoff wavelength

$$\lambda_c^{TE_{10}} = 2\,a. \tag{8.9}$$

For the critical wavelength of so-called higher modes, $\lambda_c < \lambda_c^{TE_{10}}$. The dimensions of rectangular waveguides are normalized with dimensions $a > 2b$ for practical reasons. TE_{20} with $\lambda_c^{TE_{20}} = a$ is the nearest higher mode after TE_{10}. Then follow the modes TE_{01} with $\lambda_c^{TE_{01}} = 2b$, TE_{11}, and TM_{11}, etc. The band between the dominant TE_{10} mode and the following TE_{20} mode is called the single-mode band. This band is preferred when operating the waveguide. In this band, only the dominant mode TE_{10} propagates along the waveguide. The wavelength λ of the excitation signal in the single-mode band satisfies

$$\lambda_c^{TE_{20}} < \lambda < \lambda_c^{TE_{10}} \text{ or } a < \lambda < 2a. \tag{8.10}$$

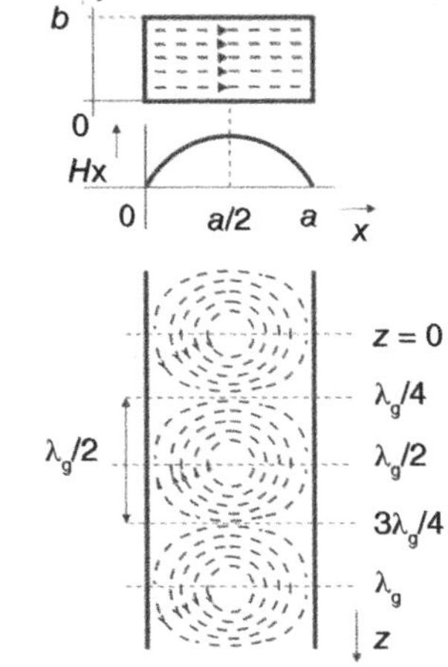

Figure 8.3 Electric and magnetic fields of the TE_{10} mode in a rectangular waveguide.

Equation (8.10) shows that the single-mode band is given by the ratio of wavelengths 1:2. In practice, a narrower band can be used:

$$1.1a < \lambda < 1.6\,a, \tag{8.11}$$

with a ratio of approximately 1.5.

The attenuation factor of the dominant mode TE_{10} in a rectangular waveguide[3] in neper per meter ($\mathrm{Np\,m^{-1}}$) can be calculated from

$$\alpha_{\mathrm{TE}_{10}} = \frac{R_{\mathrm{s}}}{\sqrt{\frac{\mu}{\varepsilon}}} \frac{\frac{1}{b} + \frac{2}{a}\left(\frac{f_{\mathrm{c}}}{f}\right)^2}{\sqrt{1 - \left(\frac{f_{\mathrm{c}}}{f}\right)^2}}, \tag{8.12}$$

where R_s is the high-frequency surface resistance given by Eq. (8.6). The relation shows that the attenuation of the TE_{10} mode decreases with increasing dimension b of the waveguide. At the same time, the maximum transmitted power in CW mode increases as well [5]

$$P_{\max} = \frac{1}{4}\frac{\sqrt{1 - \frac{\lambda}{\lambda_{\mathrm{c}}}}}{Z_0} E_0^2 \cdot a \cdot b. \tag{8.13}$$

Here, E_0 is the breakdown intensity of the electric field in dielectrics in the waveguide. The dimension b cannot be increased without limits since the single-mode band is narrowed that way. In practice, we select $b < a/2$.

8.1.1 Reactance Elements in a Rectangular Waveguide

The inductive iris shown in Figure 8.4 mainly causes the deformation of the magnetic field in the waveguide. Therefore, the equivalent circuit contains an inductance connected in parallel in the waveguide. For the normalized susceptance, we can write

$$b_{\mathrm{L}} = -\frac{\lambda_{\mathrm{g}}}{a} \cot^2 \frac{\pi d_{\mathrm{L}}}{2a}. \tag{8.14}$$

A capacitive iris reduces the height of the waveguide. That way, the electric field intensity is increased at a given location. The equivalent circuit therefore contains a parallel capacitor with an approximate value of the normalized susceptance

$$b_{\mathrm{C}} = -\frac{4b}{\lambda_{\mathrm{g}}} \ln \csc \frac{\pi d_{\mathrm{C}}}{2b}. \tag{8.15}$$

3 This attenuation is caused by imperfectly conductive walls of the waveguide.

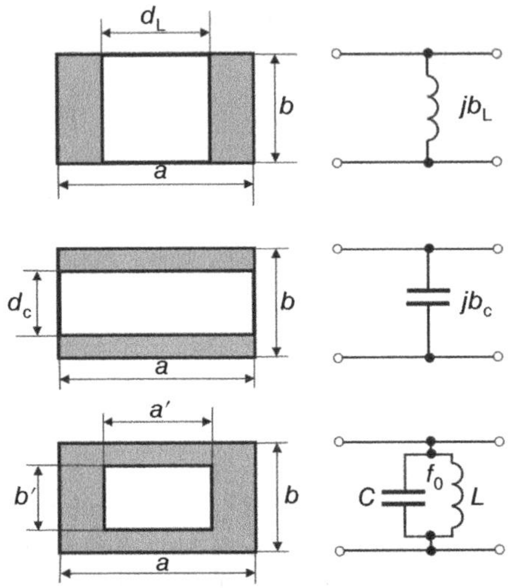

Figure 8.4 Irises in the waveguide.

A resonant iris consists of a conductive barrier with a rectangular or circular window. The barrier includes both an inductive and a capacitive iris. The equivalent circuit is a parallel resonant circuit at a given point in the waveguide. Barrier dimensions can be designed so that the dominant mode TE_{10} at a given frequency is not affected. The approximate value of the resonant wavelength λ_0 can be determined from the condition

$$\frac{b}{a\sqrt{1-\left(\frac{\lambda_0}{2a}\right)^2}} \approx \frac{b'}{a'\sqrt{1-\left(\frac{\lambda_0}{2a'}\right)^2}}. \tag{8.16}$$

Several irises with different dimensions a', b' meet the condition. The dimensions can especially influence the operating quality factor of the equivalent resonant circuit Q. The Q value is increasing with decreasing dimensions a', b'. Thus, the passband is reduced.

Let us consider two rectangular waveguides of different dimensions that are connected to each other (see Fig. 8.5). The following approximate relation holds

$$\frac{Z_{01}}{Z_{02}} \approx \frac{a_2}{a_1}\frac{b_1}{b_2}\frac{\lambda_{g1}}{\lambda_{g2}}. \tag{8.17}$$

In the equivalent circuit, a capacitive reactance appears at the interface of these waveguides.

Often used reactance elements are pins. If a conductive pin is inserted into the waveguide through the wider side (see Fig. 8.6), a reactance is created at this point (see the equivalent circuit in the same figure). The conductive part of the pin represents an inductive component of the impedance. The fringing electric field

Figure 8.5 Two coupled waveguides.

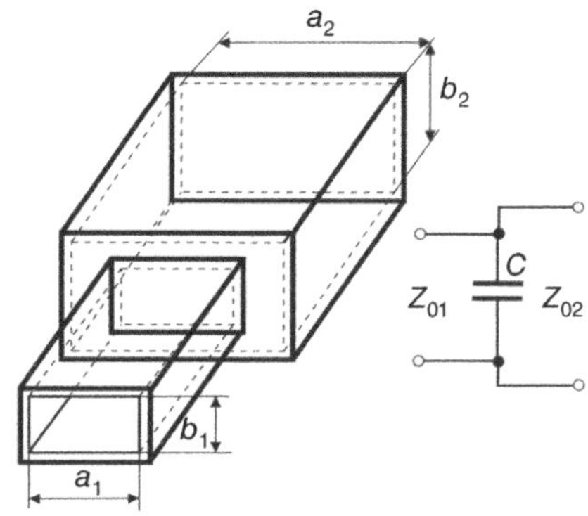

Figure 8.6 Pin in the waveguide.

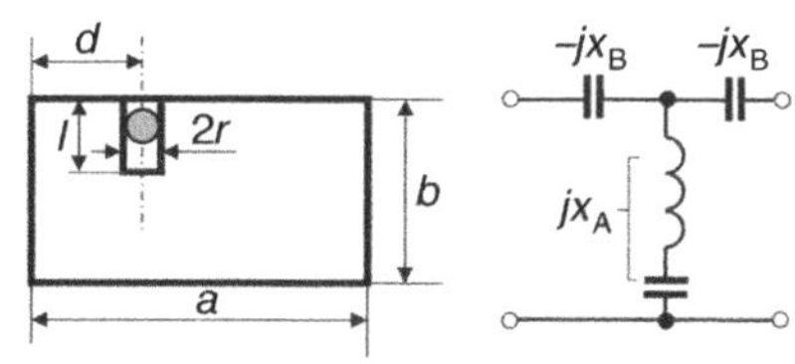

between the pin's end and the bottom wall of the waveguide represents a capacitive component. In the equivalent circuit, a series resonant circuit is thus connected in parallel to the transmission line. Longitudinal capacitive reactance is caused by the nonzero thickness of the pin. The thinner the pin, the smaller is the capacitive reactance. A series resonant circuit comes into resonance at the insertion length l_0

$$l_0 \approx \frac{\lambda_g}{4} \frac{2d - r}{2 \cdot \ln(2d/r)}. \tag{8.18}$$

That way, in practice, a short circuit is created at the given point of the waveguide. If the insertion is smaller than l_0, the pin shows capacitive reactance. If the insertion is deeper, the inductive reactance prevails. The disadvantage of such a pin is the reduction of the maximum transmitted power (the breakdown voltage is reduced).

Together with irises, pins such as tuning screws are used to construct waveguide filters of various types. Filters constructed that way can be very selective with minimal insertion loss in the passband and can transmit significant power. Two examples of waveguide bandpass filters are shown in Figure 8.7. The properties of quarter-wavelength transformers (so-called impedance inverters) are used in their construction. In the upper part of the figure, parallel resonant circuits are realized by half-wave resonators limited by inductive irises and tuned by capacitive pins (screws). The longitudinal series resonant circuit is formed by the same resonator, but is inversely transformed by two quarter-wavelength sections of the

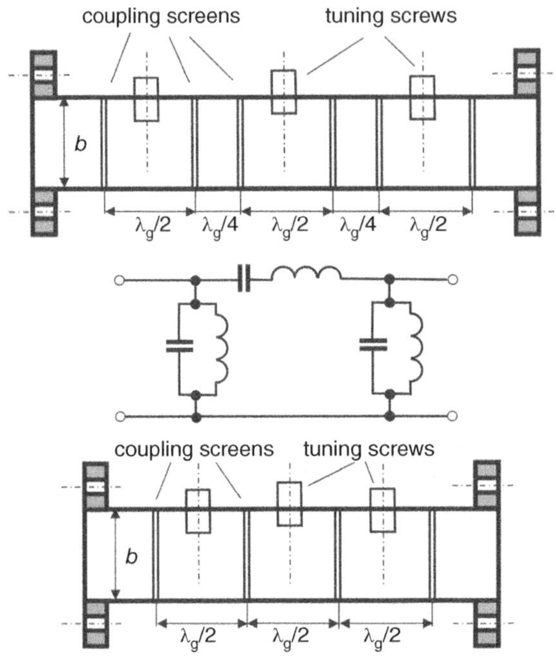

Figure 8.7 Bandpass filter in the waveguide.

waveguide. A similar solution with three directly coupled half-wavelength resonators is in the bottom part of the figure.

8.2 Circular Waveguides

Similarly to the rectangular waveguide, there is an infinite number of TE and TM waves even in the circular waveguide (see Fig. 8.8). Modes are distinguished by mode numbers m, n. The mode number m can equal to zero, but the mode number n has to be nonzero. The cutoff frequencies and wavelengths of TM modes of the circular waveguide are as follows

$$f_{\mathrm{c}}^{\mathrm{TM}} = \frac{1}{2\pi\sqrt{\varepsilon\mu}}\frac{\alpha_{mn}}{a}, \quad \text{resp.} \quad \lambda_{\mathrm{c}}^{\mathrm{TM}} = \frac{2\pi a}{\alpha_{mn}}, \tag{8.19}$$

and for TE modes,

$$f_{\mathrm{c}}^{\mathrm{TE}} = \frac{1}{2\pi\sqrt{\varepsilon\mu}}\frac{\alpha'_{mn}}{a}, \quad \lambda_{\mathrm{c}}^{\mathrm{TE}} = \frac{2\pi a}{\alpha'_{mn}}. \tag{8.20}$$

Here, α_{mn} is the n-th root of the Bessel function of the first kind and the m-th order, and α'_{mn} is the n-th root of the derivative of the Bessel function of the first

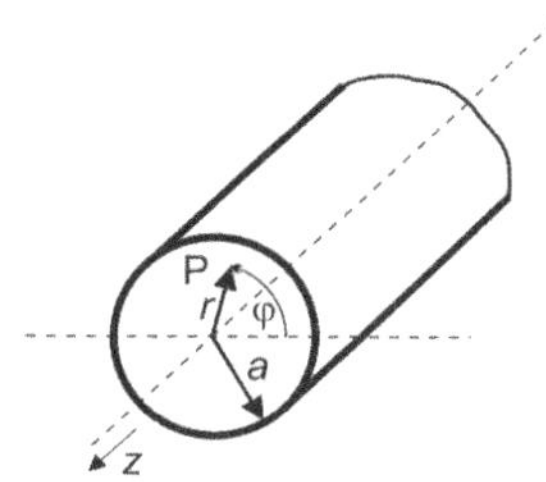

Figure 8.8 Circular waveguide.

Table 8.1 Roots of the Bessel functions of the first kind and their derivatives.

	$n = 1$	
	α_{mn}	α'_{mn}
$m = 0$	2.4048	3.8317
$m = 1$	3.8317	1.8412
$m = 2$	5.1356	3.0542
$m = 3$	6.3802	4.2012

kind and the m-th order. Table 8.1 gives roots for low values of m and $n = 1$ as an example.

The dominant mode has the lowest cutoff frequency as with the rectangular waveguide. The table shows that the root has the lowest value α'_{11}. Hence, the dominant mode of the circular waveguide is TE_{11} (see Fig. 8.9), the mode with the cutoff wavelength

$$\lambda_c^{TE_{11}} = \frac{2\pi a}{1.8412} = 3.41\, a. \tag{8.21}$$

The root α_{01} has the nearest higher value. The corresponding TM_{01} mode has the critical wavelength

$$\lambda_c^{TM_{01}} = \frac{2\pi a}{2.4048} = 2.61\, a. \tag{8.22}$$

The single-mode band of the circular waveguide is thus limited by these values.

$$\lambda_c^{TE_{11}} < \lambda < \lambda_c^{TM_{01}} \quad \text{or} \quad 2.61\, a < \lambda < 3.41\, a. \tag{8.23}$$

Obviously, compared to the rectangular waveguide, the band of the circular waveguide is narrower, and the practical limits are

$$2.72\, a < \lambda < 3.2\, a, \tag{8.24}$$

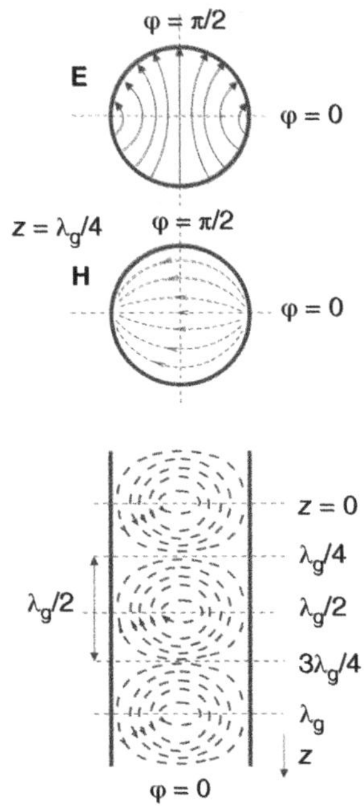

Figure 8.9 Dominant mode TE_{11} in circular waveguide.

i.e. in the ratio 1.1:1. The wavelength λ_g in a circular waveguide is given by Eq. (8.4). The attenuation factor of the dominant mode TE_{11} in neper per meter ($Np\,m^{-1}$) is given by

$$\alpha_{TE_{11}} = \frac{R_s}{\sqrt{\frac{\mu}{\varepsilon}}} \frac{\left(\frac{f_c}{f}\right)^2 + 0.418}{\sqrt{1 - \left(\frac{f_c}{f}\right)^2}}, \tag{8.25}$$

where the high-frequency surface resistance R_s is given by Eq. (8.6) and the cutoff frequency f_c by Eq. (8.20).

8.3 Coaxial Transmission Lines

A coaxial transmission line is a special type of a hollow metal waveguide with two insulated axisymmetric conductors in a cross-section. The TEM wave (Fig. 8.10) is the dominant mode here. The electric and magnetic field lines are in the transverse plane only, and are of the same shape both for high-frequency and direct-current signals. The TEM mode has zero cutoff frequency

$$f_c^{TEM} = 0, \lambda_c^{TEM} = \infty. \tag{8.26}$$

Figure 8.10 Structure of the coaxial line and field lines of the TEM mode.

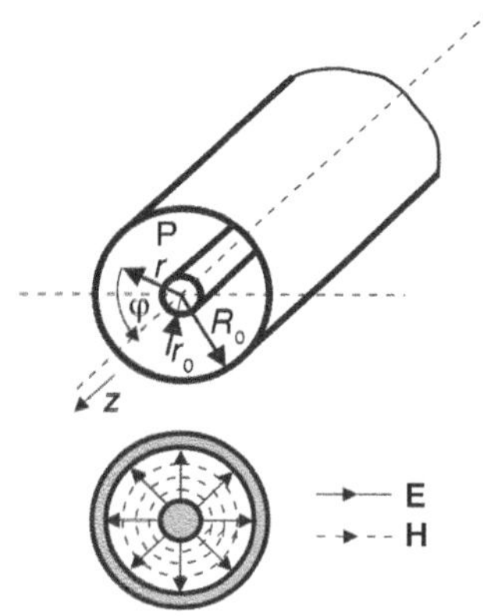

Hence, even a direct current signal with $f = 0$ can be propagated through the coaxial line. When substituting $f_c = 0$, all relations from the beginning of this chapter can be used to calculate the basic parameters of the coaxial line.

The characteristic impedance of a coaxial line is given by the ratio of complex amplitudes of voltage U between conductors and current I flowing through a conductor

$$Z_0 = \frac{U}{I} = \frac{60}{\sqrt{\varepsilon_r}} \ln \frac{R_0}{r_0}. \tag{8.27}$$

Here, ε_r is the relative permittivity of dielectrics between the conductors.

The maximum active power that can be transmitted by a TEM wave in a coaxial line is

$$P_{\max}^{\mathrm{TEM}} = \pi r_0^2 \sqrt{\frac{\varepsilon}{\mu}} E_{\max}^2 \ln \frac{R_0}{r_0}. \tag{8.28}$$

Here, $E_{\max}$ is the maximum intensity of the electric field on the surface of the inner conductor; the value must not exceed the electric strength of the dielectrics. The relation Eq. (8.28) has an extremum for $R_0/r_0 = 1.65$. Therefore, the maximum power can be transmitted by the line with impedance $Z_0\sqrt{\varepsilon_r} = 30\ \Omega$.

The attenuation factor in neper per meter (Np m^{-1}) caused by losses in conductors can be calculated from

$$\alpha_{\mathrm{TEM}} = \frac{R_s}{\sqrt{\frac{\mu}{\varepsilon}}} \frac{1}{2R_0} \frac{1 + \frac{R_0}{r_0}}{\ln \frac{R_0}{r_0}}. \tag{8.29}$$

Here, R_s is the high-frequency surface resistance of both conductors, given by Eq. (8.6). In this case, the minimum attenuation is achieved at the ratio $R_0/r_0 = 3.6$. Substituting into Eq. (8.27), we get $Z_0\sqrt{\varepsilon_r} = 77\ \Omega$. For frequently used dielectrics Teflon and polyethylene (relative permittivity $\varepsilon_r \approx 2.2$ to 2.4), the minimum attenuation is achieved for $Z_0 = 52\,\Omega$ or $50\,\Omega$.

Along the axisymmetric transmission line, TE and TM waveguide modes can also propagate next to the TEM wave. However, their use is quite exceptional. The TE_{11} mode with the largest cutoff wavelength is the main waveguide mode

$$\lambda_c^{TE_{11}} = \pi\,(R_0 + r_0). \tag{8.30}$$

In fact, this is the length of the circumference of the circle at the center of the dielectrics in a cross-section. This main waveguide mode and the higher modes are not excited in the coaxial line if the following inequality is satisfied

$$\lambda > \lambda_c^{TE_{11}} = \pi\,(R_0 + r_0). \tag{8.31}$$

That way, the single-mode band of the coaxial line is limited.

8.4 Planar Transmission Lines

Current microwave technologies extensively use planar structures. Planar structures can be implemented similarly to printed circuit boards used at lower frequencies. Those structures are the basis of hybrid MMICs as well. There are many planar structures. Microstrip transmission line, coplanar waveguide, and grounded coplanar waveguide are among the most important ones [6].

Microstrip transmission lines (see Fig. 8.11) are frequently used, especially at lower microwave frequencies. An exact analysis of this line is very difficult and can be done numerically only. There is a significant fringing field around the microstrip, which can be affected by the necessary shielding. The transverse inhomogeneity of the line (part of the signal propagates in the air and part in dielectrics with relative permittivity ε_r) causes dispersion. The wave is called HEM (hybrid electromagnetic) wave. There is no analytical solution of such a structure. Varying the width of the microstrip w (e.g. when calculating the

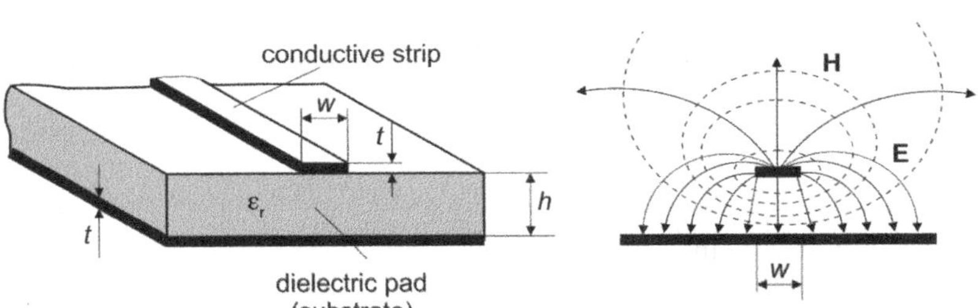

Figure 8.11 Microstrip transmission line and electromagnetic field distribution of the quasi-TEM wave.

characteristic impedance), the ratio between the signal propagating in the air and in dielectrics is changed as well. For calculations, the HEM wave is replaced by a quasi-TEM wave. Using a conformal mapping, the so-called effective parameters $\varepsilon_{\text{eff r}}$ and w_{eff} are evaluated by numerical methods. The characteristic impedance is given then by

$$Z_0 = \frac{120\pi}{\sqrt{\varepsilon_{\text{eff r}}}} \frac{h}{w_{\text{eff}}}. \tag{8.32}$$

where for a narrow microstrip with $w/h \leq 1$,

$$\varepsilon_{\text{eff r}} = \frac{\varepsilon_{\text{r}} + 1}{2} + \frac{0.9}{\pi} \frac{\varepsilon_{\text{r}} - 1}{\ln \frac{8h}{w}}, \tag{8.33}$$

$$w_{\text{eff}} = \frac{2\pi h}{\ln\left(\frac{8h}{w} + \frac{w}{4h}\right)}, \tag{8.34}$$

or for a microstrip $w/h \geq 1$

$$\varepsilon_{\text{eff r}} = \varepsilon_{\text{r}} - \frac{\varepsilon_{\text{r}} - 1}{2} \frac{\ln\left[6.28\left(\frac{w}{2h} + 0.85\right)\right]}{\frac{w}{h} + \frac{2}{\pi}\ln\left[17.08\left(\frac{w}{2h} + 0.85\right)\right]}, \tag{8.35}$$

$$w_{\text{eff}} = w + \frac{2h}{\pi} \ln\left[17.08\left(\frac{w}{2h} + 0.85\right)\right]. \tag{8.36}$$

Values of the effective relative permittivity $\varepsilon_{\text{eff r}}$ can vary within $(\varepsilon_{\text{r}} + 1)/2 \leq \varepsilon_{\text{eff r}} \leq \varepsilon_{\text{r}}$.

In the previous calculations, the nonzero thickness of the microstrip was compensated by the expansion of w on w' according to the relation

$$w' = w + \Delta w, \tag{8.37}$$

where

$$\Delta w = \frac{t}{\pi}\left(\ln\frac{2h}{t} + 1\right) \text{ for } w/h \geq 1/(2\pi) \tag{8.38}$$

or

$$\Delta w = \frac{t}{\pi}\left(\ln\frac{4\pi w}{t} + 1\right) \text{ for } w/h \leq 1/(2\pi). \tag{8.39}$$

Also the phase velocity, the group velocity, and the wavelength of the quasi-TEM wave depend on the dimensions w and h

$$v_{\text{p}} = v_{\tau} = \frac{1}{\sqrt{\varepsilon_{\text{eff}}\,\mu}} = \frac{c}{\sqrt{\varepsilon_{\text{eff r}}}} \tag{8.40}$$

and

$$\lambda_{\text{g}} = \frac{v_{\text{p}}}{f}. \tag{8.41}$$

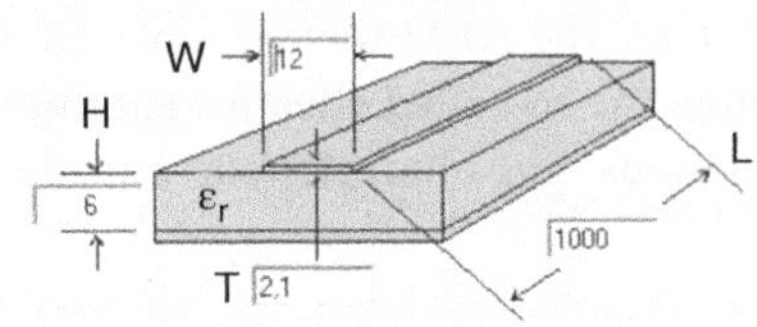

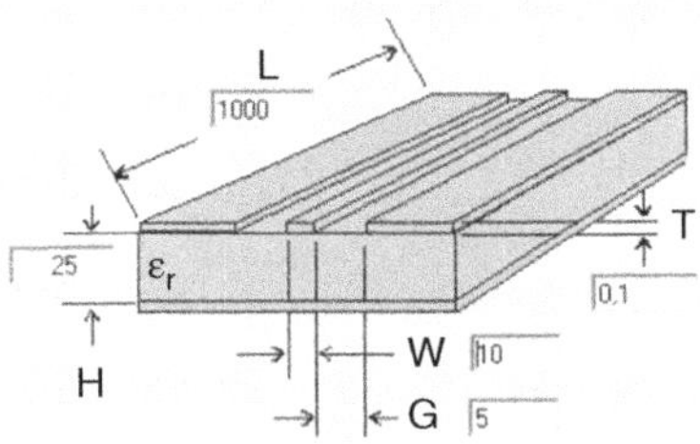

Figure 8.12 Models of microstrip transmission line and coplanar waveguide in the program AppCAD. Dimensions are in mils.

In the microstrip transmission line, the single-mode band of the quasi-TEM mode is limited by the excitation of the TE_{10} mode

$$f < f_c^{TE_{10}} = \frac{Z_0}{2\mu_0 h}. \tag{8.42}$$

In a microstrip transmission line, even a surface wave with the lowest mode TE_1 can be excited. The single-mode band is limited then from above

$$f < f_c^{TE_1} = \frac{c}{4h\sqrt{\varepsilon_r - 1}}. \tag{8.43}$$

At frequencies higher than approximately 6 GHz, dispersion begins to apply in asymmetric microstrip lines, and the above parameters become frequency dependent. Correction formulas for given frequencies have to be used then in calculations. Fortunately, there are many computer applications that can perform sufficiently accurate calculations. The module for the analysis of planar structures is a part of most programs for simulation of electromagnetic fields and radio frequency circuits. There are even small, freely available applications such as AppCAD [7]. This applies to other planar structures such as coplanar waveguides and coupled lines (see Fig. 8.12).

8.5 Nonreciprocal Microwave Lines

There are many nonreciprocal microwave structures. A ferrite circulator (see Fig. 8.13) is a typical representative. This circulator is based on electromagnetic wave propagation in a gyromagnetic environment. A wave entering

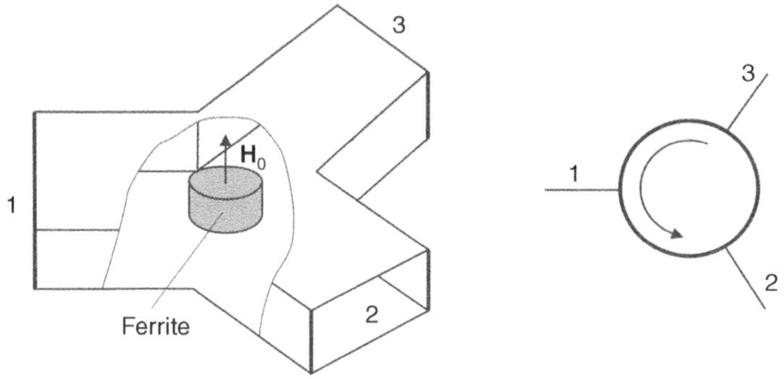

Figure 8.13 Waveguide ferrite circulator.

arm 1 exits arm 2. Similarly, a wave entering arm 2 exits arm 3 (in the direction of the arrow). Transmission in the direction opposite to the arrow is not possible. This star circulator is the most widespread, and can be implemented by waveguide, coaxial, and microstrip technology (see Fig. 8.14). The ferrite inside the resonator is a kind of a cylindrical cavity resonator. If the magnetic field H_0 excited by the ferrite equals zero, then the wave entering the arm 1 is equally split into arms 2 and 3. Applying a DC magnetic field of an appropriate intensity and direction, the high-frequency electromagnetic field is rotated so that the field intensity in the arm 3 is zero and nearly all the wave energy from arm 1 exits arm 2. This is an excellent feature that allows to separate the transmitter power stage (connected to arm 1) and the antenna (connected to arm 2), for example. The reflected wave from an imperfectly matched antenna is converted into Joule

Figure 8.14 Microstrip ferrite circulator.

heating in a matched load (connected to arm 3). That way, a high-quality isolator is created. Frequently, an isolator with two ports only can be met since the matched load (connected to port 3), is already inside the body of the isolator. In case of power applications, the load should be outside of the circulator and should be cooled for the corresponding power. The isolator can be used for the separation of individual stages of a transmitter to increase stability. Circulators

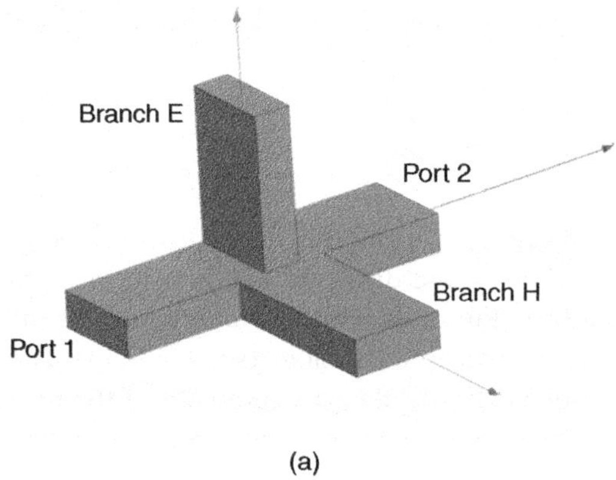

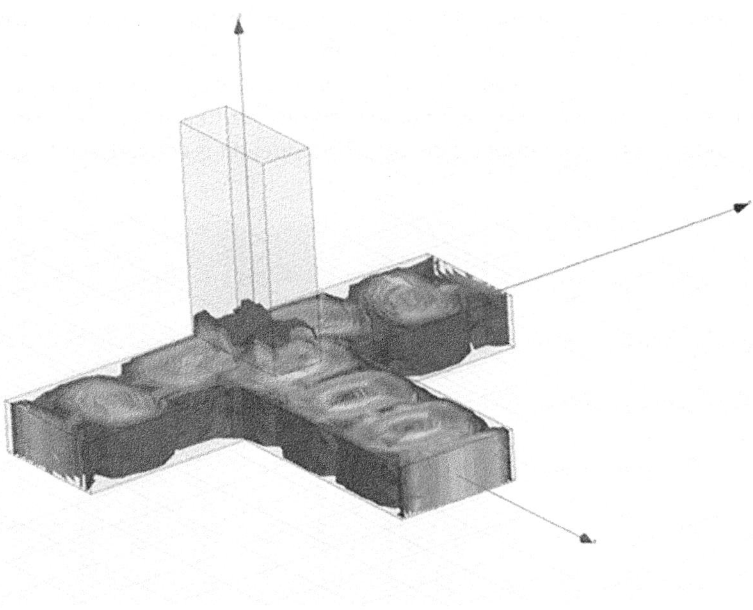

Figure 8.15 (a) Magic T; (b) In-phase excitation of magic T.

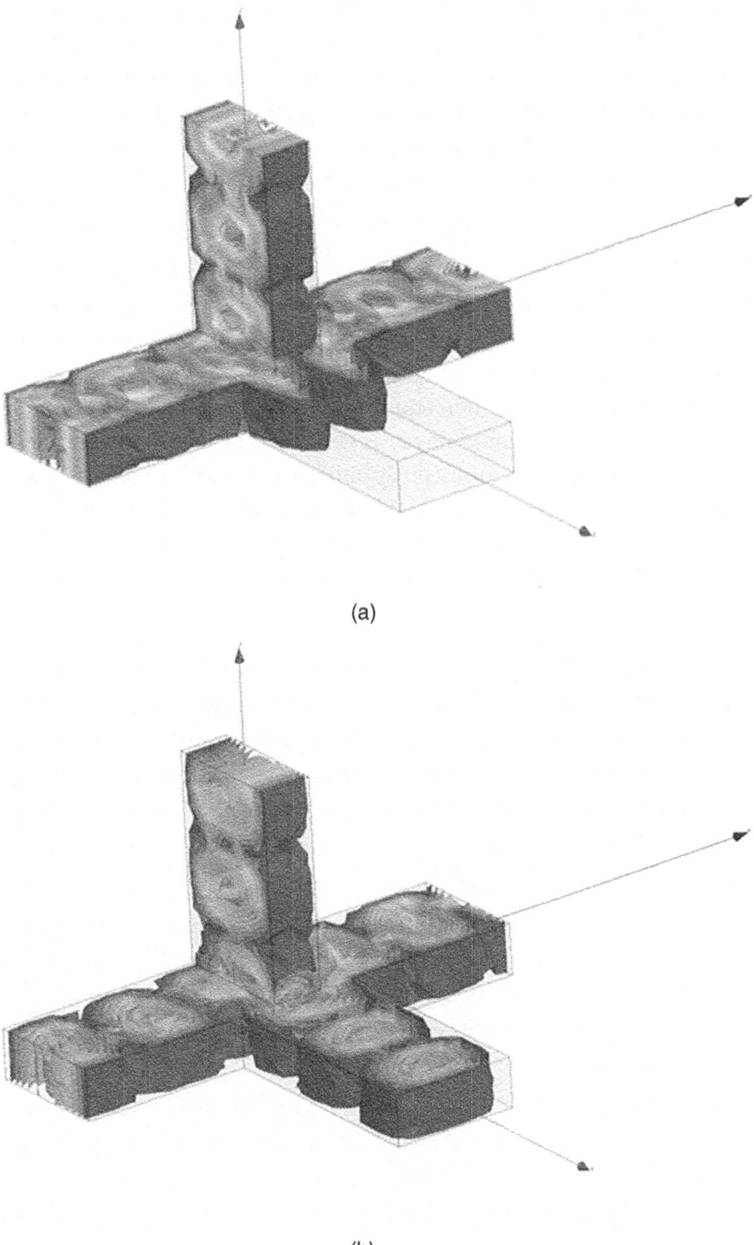

Figure 8.16 (a) Anti-phase excitation of magic T; (b) Quadrature excitation of magic T.

can be cascaded to separate several transmitters and receivers sharing a single antenna.

The magic T is another type of nonreciprocal line. It is a waveguide structure designed to combine two signals efficiently, or to split the signal into two paths. Figure 8.15a shows the fundamental structure of the magic T. The behavior of the four-port is strange at first sight (hence the name "magic"). If the same TE_{10} wave enters ports 1 and 2 in phase, the sum signal gets out of the H branch, as shown in Figure 8.15b. The signal in branch E is given by the amplitude and phase asymmetry of input signals. Branch E is terminated by a non-reflective load. If TE_{10} waves in excitation ports 1 and 2 are anti-phase, the sum signal gets out of the branch E, and the branch H is terminated by a non-reflective load (see Fig. 8.16a). Figure 8.16b shows the simulation of quadrature excitation of input ports (with 90° phase shift). The output signal is then equally divided into the E and H branches.

In microwave bands, the required power is generated by combining very often. Especially, the anti-phase excitation (push-pull) is suitable from the viewpoint of even harmonics suppression. A detailed attention is therefore turned to this option. Simulation outputs of an idealized magic T with anti-phase excitation are shown in Figure 8.17. For the 24 GHz band, we usually use the WR-42 waveguide with a window 10.67 × 4.32 mm and the single-mode band Eq. (8.11) (1.1 × 10.67) mm < λ < (1.6 × 10.67) mm. The cutoff wavelength $\lambda_c = 1.6 \times 10.67 = 17.07$ mm and the corresponding f_c = 17.5 GHz. The frequency 24 GHz approaches the upper frequency of the single-mode band. For an ideal combiner, the parameters S_{31} and S_{32} should be −3 dB. Figure 8.17 shows that the value −3 dB is reached near the cutoff frequency, f_c, and at the frequency 24 GHz, S_{31} and S_{32} are almost 4 dB. These values were confirmed by measurements as well (see Fig. 8.18a,b).

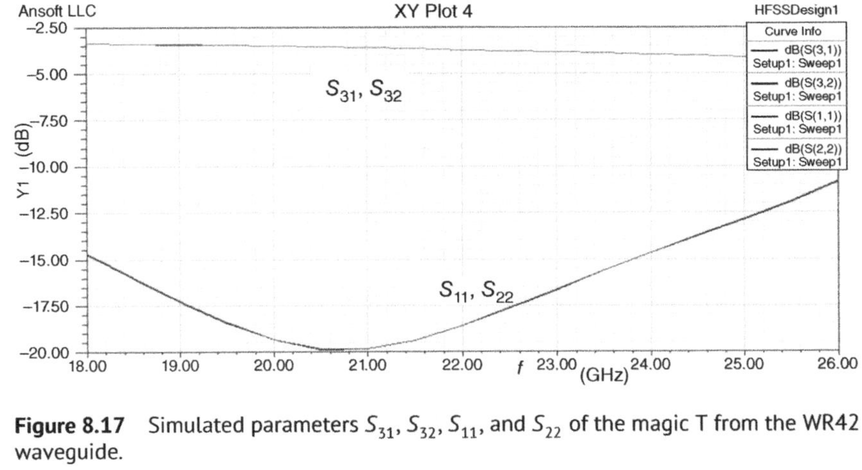

Figure 8.17 Simulated parameters S_{31}, S_{32}, S_{11}, and S_{22} of the magic T from the WR42 waveguide.

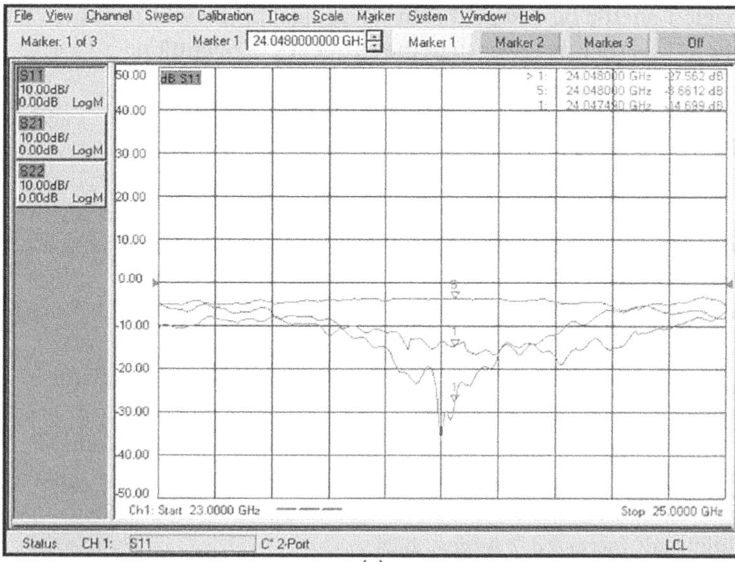

(a)

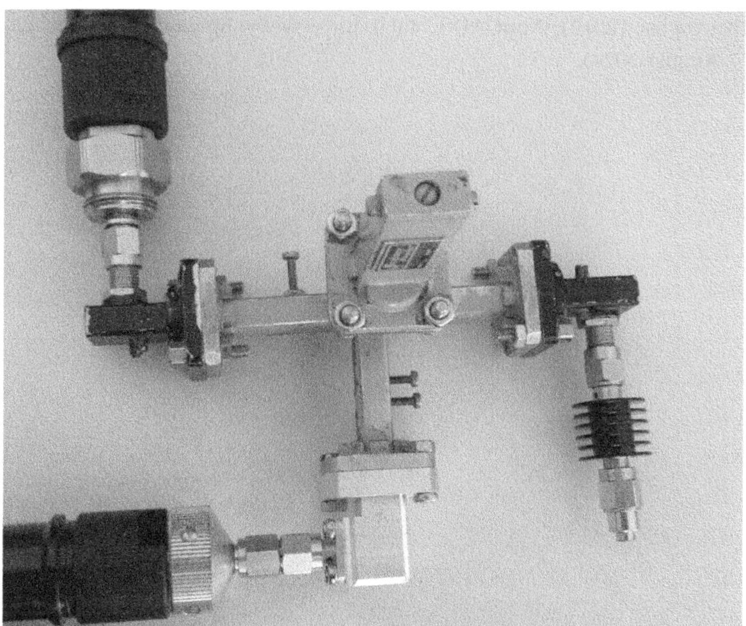

(b)

Figure 8.18 (a) Measured parameters S_{31}, S_{11}, and S_{22}, WR-42 waveguide; (b) Measurement arrangement of a magic T.

Hence, the magic T should be designed with waveguide dimensions for which the cutoff frequency, f_c, is slightly lower than the frequency of operation.

References

1 Hanus, S. and Svačina, J. (2004). *High-frequency and Microwave Technology* (in Czech). https://docplayer.cz/2727290-Vysokofrekvencni-a-mikrovlnna-technika.html (accessed 7 August 2024).

2 Tysl, V. and Růžička, V. (1989). *Theoretical Foundations of Microwave Technology* (in Czech). Prague: SNTL.

3 Kvasil, B. (1957). *Theoretical Foundations of Centimeter Wave Technology* (in Czech). Prague: SNTL.

4 Rao, R.S. (2012). *Microwave Engineering*. New Delhi: Prentice Hall of India.

5 Doebert, S. *Practical Power Capability of Rectangular Waveguides*. https://indico.cern.ch/event/245685/contributions/1565266/attachments/421127/584776/waveguidelimits.pdf (accessed 7 August 2024).

6 Balanis, C.A. (2012). *Advanced Engineering Electromagnetics*. 2nd ed. Wiley.

7 Avago Technologies. (2012). AppCAD v. 4.0.0. http://www.hp.woodshot.com/ (accessed 7 August 2024).

9

Microwave Receivers

Processing and detection of received signals are of the same importance as signal reception by an antenna. However, a fundamental difference can be found here. While the signal-to-noise ratio (SNR) can be improved by the antenna (increasing dimensions, improving reflector irradiation, etc.), the receiver can only worsen the ratio. The degradation is aimed to be as small as possible. This chapter discusses those parts and parameters of the receiver that dominantly influence the detection of very weak signals. At the input of the receiver, a low-noise amplifier (LNA) with a very low noise figure is requested. And the receiver should be designed in such a way that the resulting noise figure can approach the noise figure of the LNA itself. In microwave bands, receivers are still designed with a heterodyne. Hence, the input signal is mixed with a local oscillator signal at an intermediate frequency and is further processed and detected here, either classically (in an analog way) or sampled and further processed by digital signal processing methods. Both ways have pros and cons. Analog processing allows us to achieve higher dynamics of the receiver. Hence, a wider range of signal levels can be processed without distorting them. On the other hand, digital processing allows signals to be optimally filtered, equalized, averaged, etc. Digital receivers consist of fast A/D converters with high resolution, signal processors, field-programmable gate arrays (FPGA), and memories. These circuits, together with the frequency synthesizer of the local oscillator, form the core of a software-defined radio (SDR). By changing the program, a wide range of receiver parameters can be varied. Certainly, SDR is a concept of the future; elements of SDR have been developing and achieving better and better parameters.

Figure 9.1 shows a block diagram of the input circuitry of a typical heterodyne receiver. From the viewpoint of the best sensitivity, the input LNA has to show the lowest possible noise figure and a sufficient amplification. That way, the noise contribution of following stages is eliminated according to the Friis relation (Eq. 3.5). If desired weak signals are accompanied by strong signals at the input (even at

Reception of Weak Radio Signals from Space, First Edition. Miroslav Kasal.

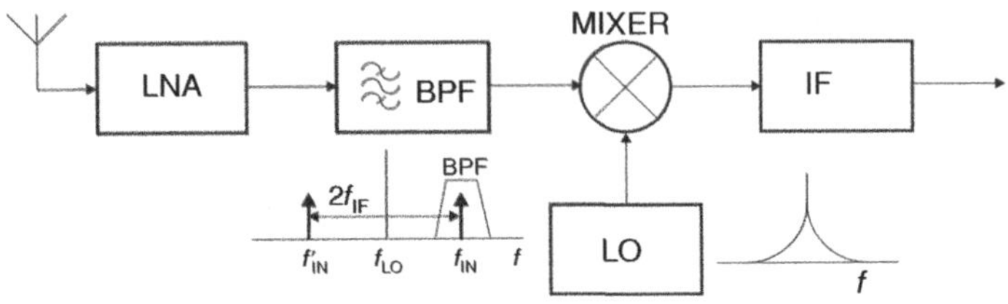

Figure 9.1 The input part of the receiver with a heterodyne.

different frequencies), the high resistance of the amplifier against those signals is very important.

The IP3, the intermodulation intercept point of the third order, is measured (see Fig. 9.2).[1] We have to distinguish if IP3 is related to the level of the input or the output. A lower IP3 is related to a higher intermodulation distortion (a generation of new components in the spectrum). Hence, noise is translated into the desired bandwidth.

A heterodyne (a mixer with a local oscillator) generates a signal with an intermediate frequency f_{IN} according to

$$f_{IN} = f_{LO} \pm f_{IF}. \tag{9.1}$$

Two input frequencies satisfy the equation—the desired one, f_{IN}, and the mirrored one, f'_{IN}. The distance in between frequencies is $2f_{IN}$ (see Fig. 9.1). The mirror frequency must be filtered out before the mixer, even if there is no signal in the mirror band. Otherwise, noise from the mirror band is translated into the intermediate frequency tract. In case of an ideal receiver with the noise figure 0 dB, this causes the degradation to 3 dB. Figure 9.3 shows the effect of mirror band suppression on the resulting noise figure. Obviously, 25–30 dB suppression of the mirror band is sufficient from the viewpoint of the translated noise.

The phase noise of a local oscillator (LO) is another source of noise in the receiver. The phase noise should be as low as possible. This condition is not always easy to satisfy (with synthesized microwave oscillators especially). This issue will be covered in more detail in Chapter 10. In the context of the current chapter, let us turn the attention to a mixer [2]. A good receiver should be double balanced—either with diodes or with unipolar FETs. The local oscillator (LO) is then better separated, and the penetration of the oscillator injection is

1 If a signal level at the input of the amplifier is gradually increased, measurable third harmonics starts to appear from a certain level. This third harmonics increases then three times faster than the fundamental harmonics. Since the third harmonics is difficult to be measured (especially at microwaves), the measurement is usually made with two signals of the same level with a small frequency separation at the input, and the third-order intermodulation product is measured.

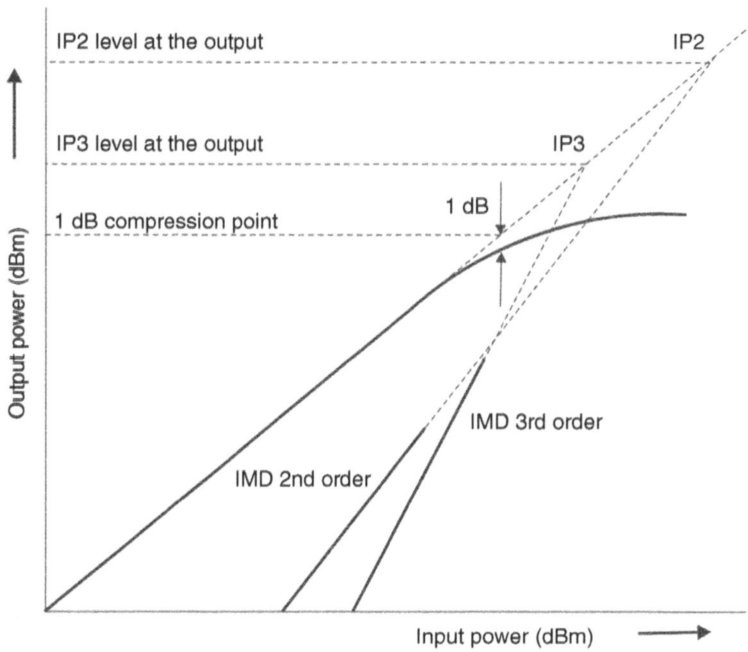

Figure 9.2 To the definition of intermodulation intersections.

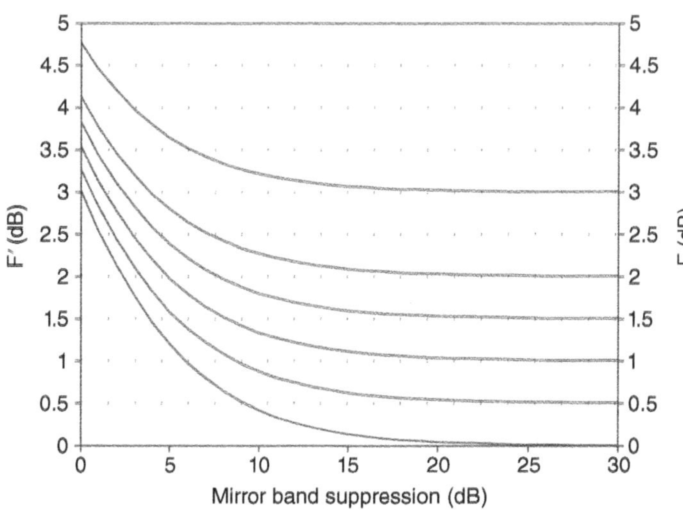

Figure 9.3 Effect of mirror band suppression on the resulting noise figure. *Source: 1. KASAL, M. (1998). Modern methods of generation and processing of nuclear magnetic resonance signals (in Czech). In: Scientific Writings of BUT, Habilitation and Inaugural Writings 1: 3–28. Brno, CR: VUTIUM. ISBN 9788021411661.*

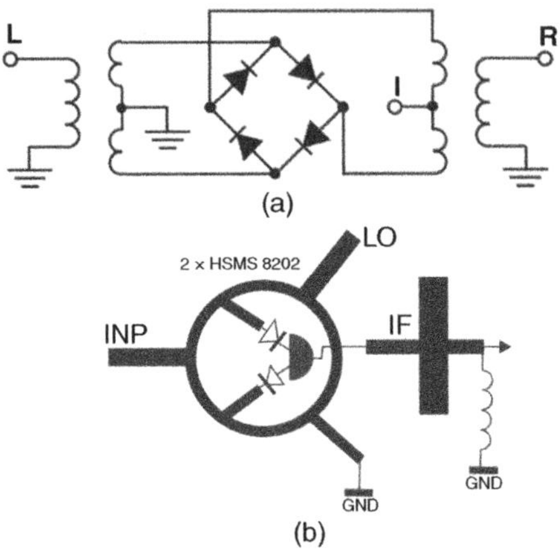

Figure 9.4 (a) Double-balanced mixer with Schottky diodes in a ring. *Source: 2. MINI CIRCUITS. (2002). RF Frequency Mixers. www.minicircuits.com/WebStore/Mixers.html (accessed 7 August 2024)*; (b) Microwave double-balanced mixer. *Source: 3. HEWLETT PACKARD. (1998). Low Cost Mixer for the 10.7 to 12.8 GHz Direct Broadcast Satellite Market. Application Note 1136. http://www.hp.woodshot.com/hprfhelp/4_downld/lit/diodelit/an1136.pdf (accessed 7 August 2024).*

more suppressed (including the "noise environment") both at the intermediate frequency and at the receiver input. At lower frequencies, high-quality double-balanced mixers (DBM) are implemented with a "ring modulator" configuration with high-frequency transformers and four ring-connected Schottky diodes (see Fig. 9.4a). The conversion slope of such a mixer is typically −6 dB and has to be considered when determining the receiver gain. At microwave frequencies, double-balanced mixers can be implemented using microstrip technology (see Fig. 9.4b, [3]). The mixer is based on a circular microstrip with an electrical length of 540°, i.e. 1.5λ (at the input frequency), with an impedance of 70 Ω. The length between sections is 90° and 180°, respectively. A grounded shunt is of the length $\lambda/4$. At the intermediate frequency output, the structure behaves as a high-frequency band-stop filter.

As already mentioned, the dynamic range of the receiver is an important parameter, and the mixer is critical from the viewpoint of this parameter. At a mixer input, there is an already amplified signal (i.e. signal of a higher level). Since the mixer is a nonlinear circuit, the second-order product is the desired signal at its output (usually). The dynamic range is defined as a ratio of the strongest signal (that can be processed by the circuit without distortion) and the weakest signal (referred to as the minimum discernible signal, MDS). Under normal circumstances, MDS is given by the equivalent thermal noise level at the mixer input. The double-balanced mixer DBM with diodes with conversion conductance −6 dB can be considered (from the viewpoint of calculating the equivalent noise temperature) as an attenuator with attenuation $L_F = 6$ dB. According to Eq. (3.7),

$$T_e = T_0(L_F - 1) = 290\left(10^{\frac{6}{10}} - 1\right) = 865 \text{ K},$$

where $T_0 = 290$ K is the reference ambient temperature. For the noise level at the mixer input, the bandwidth 3 kHz, we can write according to Eq. (1.2): $N_{MI} = k \cdot T_e \cdot B_n = 1.38 \times 10^{-23} \cdot 865 \cdot 3000 = 3.58 \times 10^{-17}$ W, so in logarithmic measure, this noise level is −164.5 dBW, or −134.5 dBm. The maximum signal level is given by the 1 dB compression level (see Fig. 9.2). For diode mixers, this level is 5–10 dB below the local oscillator level. Reputable manufacturers (e.g. Mini-Circuits) produce mixers for different applications with different LO levels ranging from +7 to +23 dBm. The achievable dynamic range is then 132–145 dB.

Contrary to the realistic 1 dB compression point, the intermodulation intersections IP2 and IP3 represent imaginary levels, based on Taylor series expansion for the second and third orders. Thanks to a precise mathematical definition, the intermodulation intersections are used as a parameter describing dynamic properties of high-frequency two-ports.

The problem of mirror band suppression can be solved with a quadrature (IQ) mixer as well. The corresponding block diagram is shown in Figure 9.5.

The quadrature mixer is based on two balanced mixers with local oscillators in a quadrature, i.e. with a phase shift of 90°. The output of one mixer is 90° delayed, and outputs of both mixers are then summed. As indicated by phasors in Figure 9.5, there is only a low sideband signal at the output of the summing element (at the intermediate frequency). The upper sideband signal is suppressed.

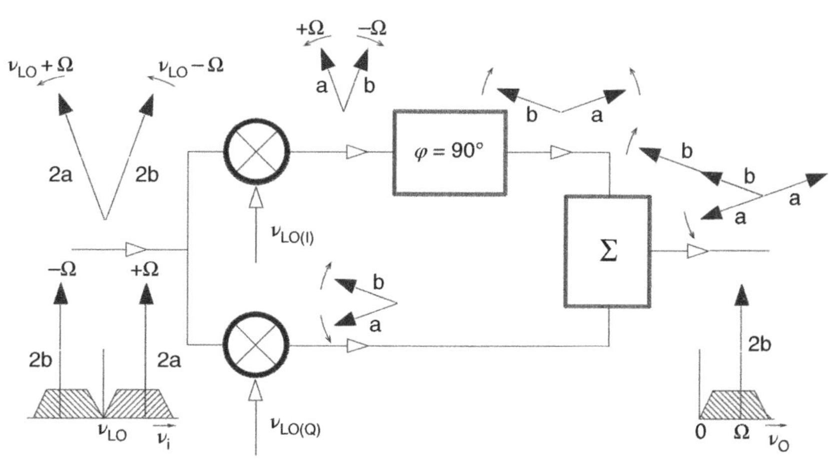

Figure 9.5 The principle of a quadrature mixer.

The IQ mixer can be implemented for nearly any intermediate frequency (including zero frequency, zero IF) which is a significant advantage.[2] Mirror frequency suppression depends on the orthogonality and amplitude symmetry of signals. If the error of quadrature is 1°, the mirror frequency is suppressed for about 40 dB.

The same suppression is achieved with an amplitude asymmetry of 1%. If both errors occur at the same time, the suppression decreases by another 3 dB [1]. These are the practical limits of an analog IQ mixer. The mixer can be used at frequencies up to 26 GHz.

Today, the concept of the IQ mixer is used in digital systems as well. The local oscillator is created by direct digital synthesis with two digital outputs with a precise 90° phase shift. "Balanced" mixers can be implemented with fast digital multipliers with perfect symmetry. A typical 80 dB suppression of mirror frequency can be achieved in such systems. So far, balanced mixers are used for secondary mixing at medium frequencies, but the trend is moving toward higher frequencies.

9.1 Low-Noise Amplifiers

Modern GaAs transistors with high mobility of charge carriers (HEMT, high-electron-mobility transistor) allow to achieve incredibly low noise figures of input amplifiers even at very high microwave frequencies. However, even here applies that "you can't make a good amplifier with a bad transistor, but you can make a bad amplifier with a good transistor."

The LNA design is usually started by considering the necessary gain, the needed immunity against strong signals, and the requested noise figure to be achieved for a given application. To determine the total noise figure related to parameters of separate stages, the Friis relation is used (Eq. 3.5). In order to prevent generation of interfering intermodulation products, a filter has to be used at the LNA input in some applications. The insertion loss of this filter acts as an attenuator at the LNA input, and worsens the LNA noise figure by the value of this attenuation (in dB). Usually, a resonator with quality factor without load Q_0 is used as such a filter. The resonator is loaded by an antenna on one side and by the first LNA transistor on the other side. The bandwidth B, i.e. the desired selectivity, is then determined by the resulting quality factor Q according to the known relation

$$B = \frac{f}{Q}. \tag{9.2}$$

2 With a classic heterodyne, the choice of intermediate frequency is important due to the implementation of a filter before the mixer. A higher intermediate frequency makes the implementation of a suitable filter simpler.

Here, f is the frequency of operation. For efficiency of the transmission[3] η for the input resonator, the following relation can be derived

$$\eta = \left(1 - \frac{Q}{Q_0}\right)^2, \tag{9.3}$$

and an additional attenuation (dB) caused by the resonator is

$$b = -10\log\eta, \quad \text{or} \quad b = -20\log\left(1 - \frac{Q}{Q_0}\right). \tag{9.4}$$

An example: A resonator at the input of LNA for the 432 MHz band is expected to have a bandwidth of 40 MHz (for a 3 dB decrease). According to Eq. (9.2), the operating quality factor of the resonator is $Q = 10.8$. If the quality factor of the unloaded resonator is $Q_0 = 300$, the additional insertion loss of the resonator is

Table 9.1 An example of the data for noise matching (CE3512K2 HEMT transistor).

CE3512K2 N-channel HJ-FET, $V_{ds} = 2\,V$, $I_d = 10\,mA$				
		Γ_{opt}		Rn/50
Freq. GHz	F_{min} dB	Mag. (–)	Ang. (°)	-
8.0	0.28	0.554	70.9	0.136
8.5	0.29	0.531	77.5	0.124
9.0	0.30	0.508	84.1	0.112
9.5	0.32	0.480	91.3	0.104
10.0	0.34	0.452	98.5	0.096
10.5	0.36	0.429	106.1	0.079
11.0	0.37	0.405	113.6	0.062
11.5	0.40	0.383	122.0	0.053
12.0	0.42	0.361	130.4	0.044
12.5	0.44	0.335	139.2	0.040
13.0	0.46	0.309	147.9	0.036
13.5	0.48	0.305	157.5	0.033
14.0	0.49	0.301	167.0	0.030

Source: California Eastern Laboratories [4].

3 The efficiency of the transmission by the resonator was derived as a ratio of a power in matched load of the resonator and an available power of signal source (which is the maximum power the signal source can deliver to the matched load).

$b = 0.32$ dB. And the noise figure of the first LNA transistor is increased at least by this value. Why at least? The noise figure would be increased by this value with simultaneous power and noise matching of the first-stage transistor, which is often impossible. Eqs. (9.3) and (9.4) show the known fact that the quality factor of the unloaded resonator Q_0 should be as high as possible and the operational quality factor Q as low as possible to reach the lowest possible insertion loss. Hence, a compromise with the desired selectivity has to be made.

9.1.1 Noise Matching

Noise matching is related to such a value of generator immittance when the excited active element (most often a transistor) generates the minimum noise. For low-noise transistors, these data are usually a part of a catalog sheet with s-parameters [4] (see Table 9.1). Here F_{min} is the minimum noise figure achievable at the reflection coefficient of the signal source Γ_{opt}, i.e. in case of noise matching.

In terms of power matching, the optimal impedance of the signal source is the complex-conjugate of s_{11}, denoted as s_{11}^* in Figure 9.6. If these values are close to each other as in Figure 9.6, a very good noise figure can be achieved. In the figure, circles of the constant noise figure are plotted for values $F = 0.6; 0.8; 1.0$ satisfying the equation

$$F = F_{\text{min}} + 4\frac{R_{\text{n}}}{Z_0}\frac{|\Gamma_{\text{s}} - \Gamma_{\text{opt}}|^2}{|1 - \Gamma_{\text{opt}}|^2\left(1 - |\Gamma_{\text{s}}|^2\right)}. \tag{9.5}$$

Here, F and F_{min} are noise figures (–), R_{n} is the equivalent noise resistance (Ω), Z_0 denotes the reference impedance (Ω), Γ_{s} is reflection coefficient of the signal source (–), and Γ_{opt} denotes the value corresponding to the noise matching (–).

The matching should be made as simple and as lossless as possible. Figure 9.7 shows the noise matching based on two lines to a signal source with an impedance of 50 Ω.

9.1.2 Cooled Amplifiers

Amplifiers with extremely low noise used in radio astronomy and in communication with interplanetary probes, are implemented using quantum amplifiers (maser) operating at very low temperatures, approaching the temperature of liquid helium (4 K). Such a technology requires a complex and bulky cryotechnology, which cannot be placed into the focus of a parabolic reflector. The feed has to be connected to the maser by a waveguide cooled by liquid nitrogen (77 K) or by a system of auxiliary reflectors known as beam waveguide (BWG) technology.

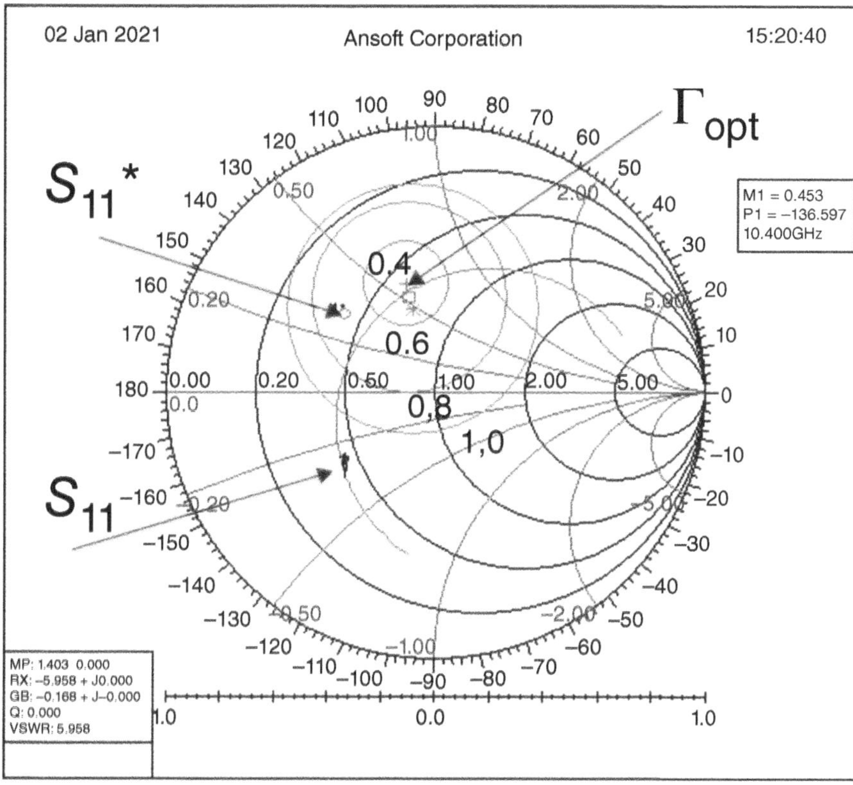

Figure 9.6 The circle of a constant noise figure at 10.4 GHz. Optimal impedance of the signal source is between values of Γ_{opt} and s_{11}^*.

Modern pHEMT or InP HFETs[4] allow a relatively simple integration of the LNA into the feed of the reflector directly. That way, the losses between the feed and the amplifier are minimized. Or the amplifier can be cooled to reduce the noise [5]. Noise characteristics of such amplifiers are approaching systems with masers. However, the design is simpler and cheaper since the cooling capacity is incomparably smaller. A substantial reduction of the noise figure can be achieved by cooling to a nitrogen temperature of 77 K, which is considerably easier than cooling with helium. An example of a cryo-amplifier made by the Swedish company Low Noise Factory [6] is shown in Figure 9.8a.

4 Bipolar transistors with a SiGe HBT heterostructure are used for the construction of microwave LNA as well.

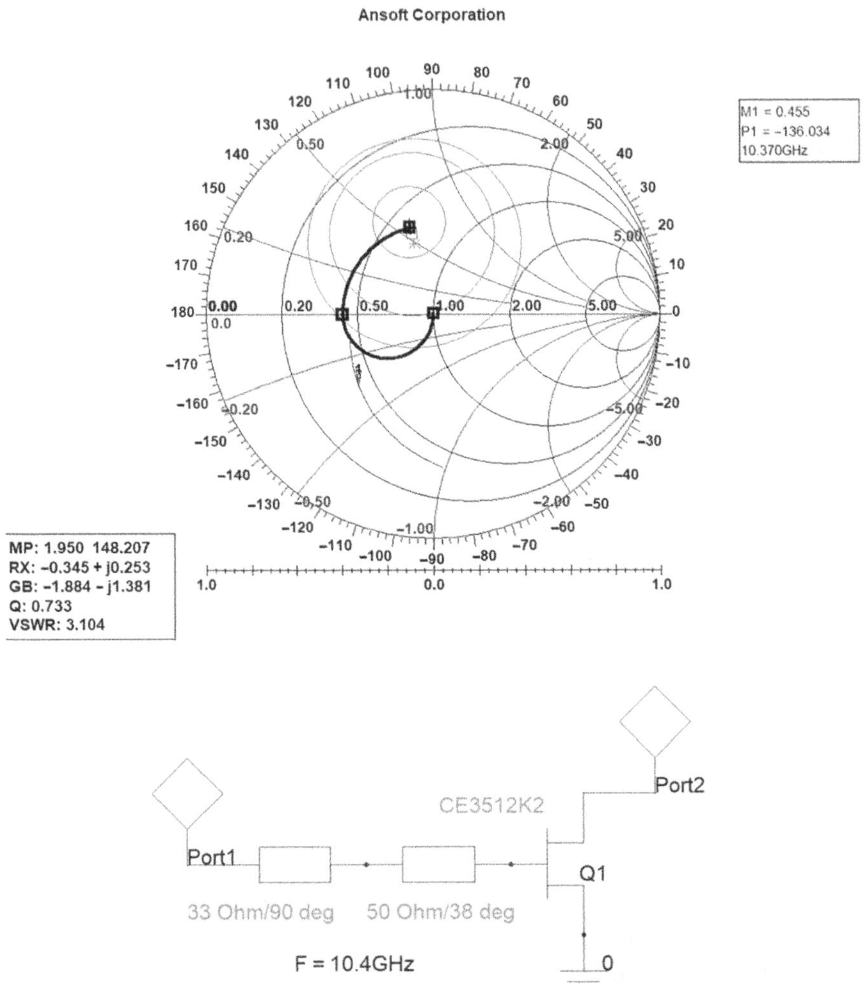

Figure 9.7 Noise matching based on two lines.

Figure 9.8b shows the measured equivalent noise temperature $T_e = 6$ K. At the reference temperature, $T_0 = 290$ K, the noise figure is given by Eq. (3.4) $F = (290 + 6)/290 = 1.02$, i.e. 0.09 dB.

The reference temperature should correspond to the temperature of the signal source. Therefore, such noise figure would be measured for an undercooled amplifier with a signal source at a temperature of 290 K. Finally, Figure 9.9 shows the measured equivalent noise temperature of the amplifier, $T_e = 80$ K, at room

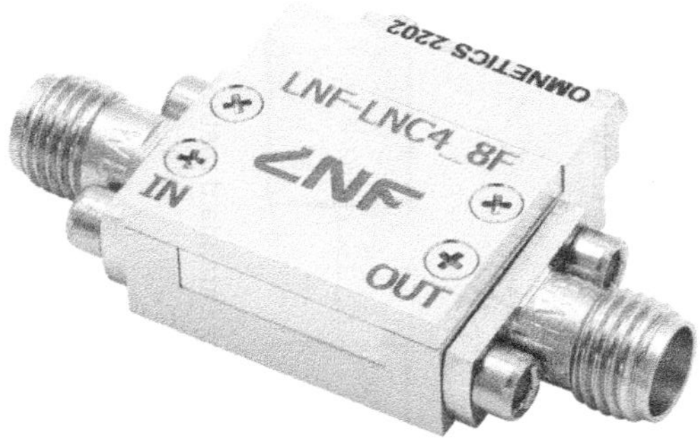

(a)

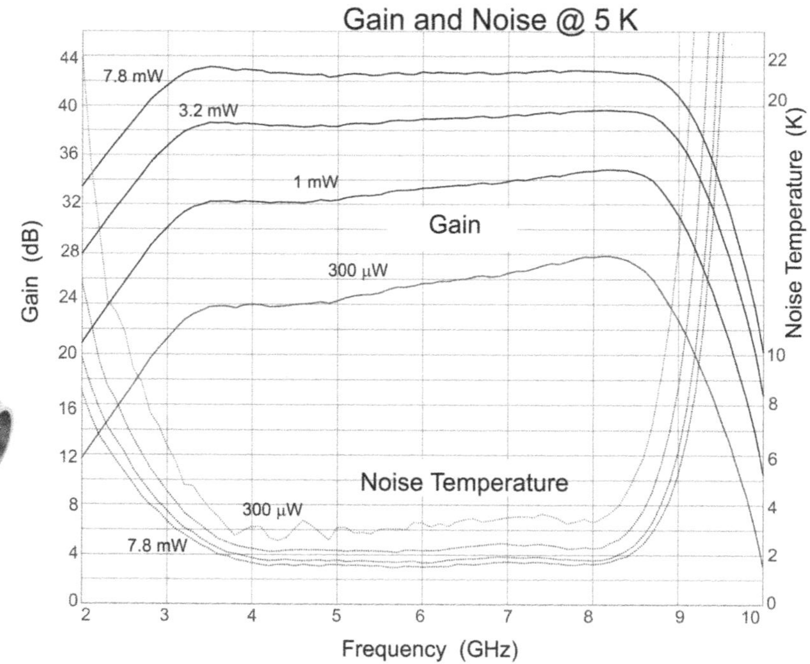

(b)

Figure 9.8 (a) Cryo-LNA for 6–11 GHz made by Low Noise Factory. *Source:* [6] / *LOW NOISE FACTORY. www.lownoisefactory.com/products/ (accessed 7 August 2024)*; (b) Gain and noise temperature response of an amplifier undercooled to 10 K.

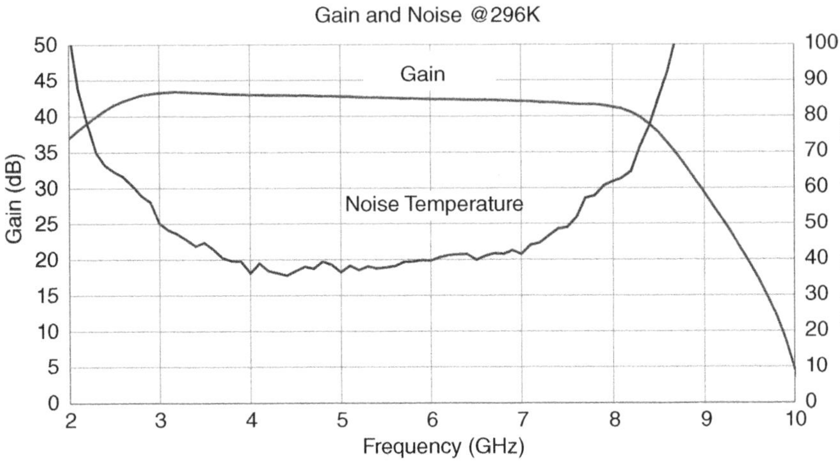

Figure 9.9 Gain and noise temperature response at a temperature of 296 K.

temperature. Hence, $F = (290 + 80)/290 = 1.27$ (1.04 dB), which is the value of the noise figure achievable at room temperature.

9.1.3 Measurement of Noise Figure

The standard method of measuring the noise figure, which is defined together with the equivalent noise temperature in Chapter 3 (3.4), is shown in Figure 9.10 [7, 8]. In order to minimize measurement errors, the method has to be applied precisely, especially when measuring very small-noise figures, less than 1 dB. The noise head can have two states characterized by temperatures T^{S}_{OFF} and T^{S}_{ON}, corresponding to switch-off and switch-on of the noise source.[5] The noise head is

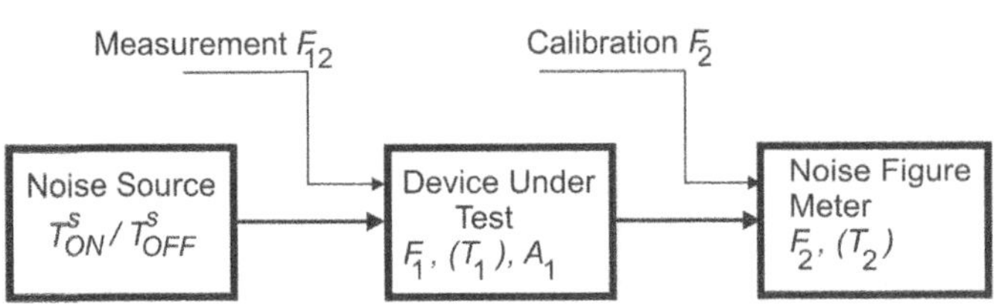

Figure 9.10 Noise figure measurement by a standard method using a noise head and a calibrated receiver—a noise figure meter.

5 Most noise heads are switched on with +28 V. The switch off corresponds to 0 V. During calibration and measurement, the noise head is supplied with a pulsed voltage with duty cycle 1:1.

precisely calibrated during production, and its basic parameter, called excess noise ratio (ENR), is given by

$$\mathrm{ENR} = \frac{T^{\mathrm{S}}_{\mathrm{ON}} - T^{\mathrm{S}}_{\mathrm{OFF}}}{T_0}, \tag{9.6}$$

where, as given by the definition of the noise figure, $T_0 = 290$ K.[6]

ENR is usually expressed in logarithmic scale (dB)

$$\mathrm{ENR} = 10 \log \left[\frac{T^{\mathrm{S}}_{\mathrm{ON}} - T^{\mathrm{S}}_{\mathrm{OFF}}}{T_0} \right] \tag{9.7}$$

and depends on the frequency of course. The accuracy of determining ENR, together with the accuracy of the noise head impedance (50 Ω), give the accuracy of the noise figure measurement.

The noise figure measurement is carried out by measuring the ratio of two noise powers by the receiver

$$Y = \frac{N_{\mathrm{ON}}}{N_{\mathrm{OFF}}}. \tag{9.8}$$

Since both noise powers are proportional to the temperature, Equation (9.8) can be written as

$$Y = \frac{T_{\mathrm{ON}}}{T_{\mathrm{OFF}}} \tag{9.9}$$

Before the measurement, the measuring receiver has to be calibrated. Hence, the equivalent noise temperature T_2 has to be determined. The noise head is directly connected to the input of the measuring receiver, and we measure

$$Y_2 = \frac{T_{2\,\mathrm{ON}}}{T_{2\,\mathrm{OFF}}} = \frac{T^{\mathrm{S}}_{\mathrm{ON}} + T_2}{T^{\mathrm{S}}_{\mathrm{OFF}} + T_2} \quad \text{and hence} \quad T_2 = \frac{T^{\mathrm{S}}_{\mathrm{ON}} - Y_2 T^{\mathrm{S}}_{\mathrm{OFF}}}{Y_2 - 1}. \tag{9.10}$$

Here, $T^{\mathrm{S}}_{\mathrm{OFF}}$ is the real thermodynamic temperature of the measuring head, and $T^{\mathrm{S}}_{\mathrm{ON}}$ is the temperature calculated from ENR (Eqs. 9.6 and 9.7). If $T^{\mathrm{S}}_{\mathrm{OFF}} = T_0$, the last relation for T_2 is equivalent to

$$F_2 = \frac{\mathrm{ENR}}{Y_2 - 1}. \tag{9.11}$$

During the calibration and measurement, the process is repeated several times, and the measured values are averaged. At the end of the calibration, the device saves the values $N_{2\,\mathrm{OFF}}$ and $N_{2\,\mathrm{ON}}$, and calculates Y_2 and T_2. Based on these calculations, the noise figure of the meter is normalized to be zero. In parallel, the gain is "zeroed" as well.

---

6 Since the noise head is designed to measure the noise figure, the head is always calibrated to $T_0 = 290$ K.

Now, the measured two-port can be inserted between the noise head and the receiver, and the measurement process can be started. The receiver is measuring

$$Y_{12} = \frac{N_{12\,\mathrm{ON}}}{N_{12\,\mathrm{OFF}}} = \frac{T_{12\,\mathrm{ON}}}{T_{12\,\mathrm{OFF}}} \tag{9.12}$$

and from there, similarly to Eq. 9.10,

$$T_{12} = \frac{T^{\mathrm{S}}_{\mathrm{ON}} - Y_{12}T^{\mathrm{S}}_{\mathrm{OFF}}}{Y_{12} - 1}. \tag{9.13}$$

Here, T_{12} is the equivalent noise temperature of the measured two-port connected to the receiver of the meter. For $T^{\mathrm{S}}_{\mathrm{OFF}} = T_0$ we can write, similarly to Eq. 9.11,

$$F_{12} = \frac{\mathrm{ENR}}{Y_{12} - 1}. \tag{9.14}$$

Since the device saved the values $N_{2\,\mathrm{OFF}}$ and $N_{2\,\mathrm{ON}}$ when calibrating and since $N_{12\,\mathrm{OFF}}$ and $N_{12\,\mathrm{ON}}$ are known, the power transfer of the measured two-port can be calculated:

$$A_1 = \frac{N_{12\,\mathrm{ON}} - N_{12\,\mathrm{OFF}}}{N_{2\,\mathrm{ON}} - N_{2\,\mathrm{OFF}}}, \tag{9.15}$$

and the correction can be done following the Friis relation (Eq. 3.6)

$$T_1 = T_{12} - \frac{T_2}{A_1}. \tag{9.16}$$

Hence, the equivalent noise temperature of the measured two-port, T_1, is known. According to Eqs. 3.4 and 3.1, the noise factor and the noise figure are, respectively,

$$F_1 = 1 + \frac{T_1}{290}, \quad \text{resp.} \quad F_{1\,\mathrm{dB}} = 10 \log F_1. \tag{9.17}$$

Even the power transmission, A_1, is known, and the calculated gain (in decibels) is

$$A_{1\,\mathrm{dB}} = 10\,\log A_1. \tag{9.18}$$

The measurement accuracy is affected by several factors:

a) The internal impedance of the noise head usually slightly varies when switched. A high-quality attenuator should be therefore connected to the output of the noise head, which suppresses those variations and reduces ENR by the value of the insertion loss of the attenuator.
b) ENR should be as small as possible, especially when measuring very small-noise figures, less than 1 dB. The receiver is important to operate in the

linear region for all measured levels. Therefore, the gain of the receiver has to be chosen according to the gain of the measured two-port.

c) The noise head is usually equipped with a precise N or SMA (2.92 mm) connector at the output. When measuring different amplifiers, different adapters to the input connector or waveguide of the amplifier have to be used. The insertion loss of these adapters L_0(dB) is directly added to the noise figure of the amplifier,

$$F_1' = F_1 + L_0 \quad \text{and hence the correction} \quad F_1 = F_1' - L_0. \tag{9.19}$$

When using adapters, attention has to be paid to the impedances. Only the highest quality connectors and adapters, and correctly set torque wrenches, should be used.

d) The larger the receiver bandwidth, the shorter the time required to average the noise level to achieve the same jitter. Typically, the measurement is performed with the bandwidth 3–4 MHz. A problem can arise if a narrower filter is part of the two-port under investigation. Modern meters with SDR receivers can narrow the bandwidth down to a few kHz at the cost of extending the averaging by the same ratio.

e) If a frequency conversion (mixer) is a part of the measured two-port, the double sideband (DSB) or single sideband (SSB) noise figure has to be clarified to be measured. If an appropriate filter is included in the two-port before the mixer (see the introductory part of this chapter), the SSB noise figure is measured in a standard way. If no filter is included, the mixer translates the noise from both sidebands into the intermediate frequency band, and the DSB noise figure is measured. The DSB is higher than the SSB noise figure by 3 dB (2x) if ENR is the same in both sidebands. When measuring a two-port with a mixer, the local oscillator of sufficient quality has to be used, and should not affect the measurement with phase noise.

f) Temperatures tend to be the most severe problem. As already shown, the noise figure is by definition related to T_0 = 290 K. If $T^{\mathrm{S}}_{\mathrm{OFF}}$ = 290 K, everything is all right. The temperature 290 K corresponds to 17 °C. However, measurements are usually performed at a higher temperature, and a correction has to be done. Let us consider the ambient temperature T_A. At this temperature,

$$\mathrm{ENR}' = \frac{T_0}{T_\mathrm{A}}\mathrm{ENR}, \quad \text{resp.} \quad ENR'_{\mathrm{dB}} = 10\log\frac{T_0}{T_\mathrm{A}} + \mathrm{ENR}_{\mathrm{dB}}. \tag{9.20}$$

At T_A = 296 K(23 °C), ENR' is for 0.089 dB lower than ENR for T_0 = 290 K.

The following corrections can be used for the measured noise figure. For measurements without calibration,

$$F'_{12} = \frac{\mathrm{ENR}}{Y-1} + 1 - \frac{T_{\mathrm{A}}}{T_0}, \tag{9.21}$$

and for measurements with calibration,

$$F'_1 = F'_{12} - \frac{F_2 - \dfrac{T_{\mathrm{A}}}{T_0}}{A_1}. \tag{9.22}$$

At $T_{\mathrm{A}} = 296$ K, the corrected values are $F'_{12} = F_{12} - 0.02$ and $F'_1 = F'_{12} - (F_2 - 1.02)/A_1$. Let us consider that $F_2 = 2$ dB (1.58), $F_{12} = 0.806$ dB (1.204), and $A_1 = 13$ dB (20) were measured at this temperature. From these values, $F_1 = 0.7$ dB (1.175). If $T_{\mathrm{A}} = T_0 = 290$ K then that would be the case. For the temperature $T_{\mathrm{A}} = 296$ K, a correction according to Eqs. (9.21) and (9.22) is done: $F'_{12} = 0.733$ dB (1.184) and $F'_1 = 0.629$ dB (1.156). These corrections are related to the noise figure measurement and not to temperature variations of the noise figure of the measured two-port. In [7], the chart in Figure 9.11 shows the measurement error depending on the ambient temperature.

Modern noise figure meters can correct the influence of current temperature T_{A} automatically. For these devices, Keysight Technology offers the Noise Figure Uncertainty Calculator [9] enabling to objectify even "miraculous" values of measured noise figures.

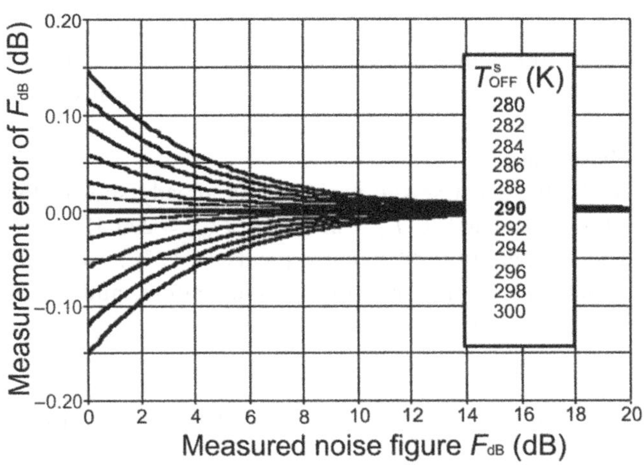

Figure 9.11 Error of noise figure measurement when $T^{\mathrm{S}}_{\mathrm{OFF}} \neq 290$ K. *Source: 7. AGILENT TECHNOLOGIES. (1988). Noise Figure Measurement Accuracy: The Y-Factor Method. Application Note AN 57-2. http://hparchive.com/Application_Notes/HP-AN-57-2.pdf (accessed 7 August 2024).*

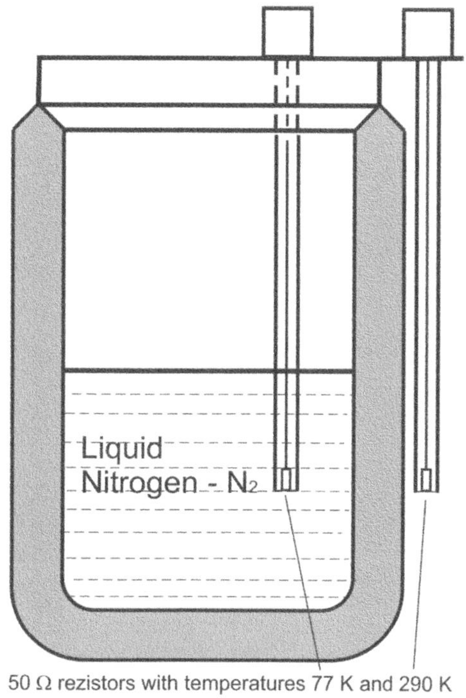

Figure 9.12 An open polystyrene container with liquid nitrogen for measuring the noise figure using resistors of different temperatures.

There are even other methods of noise figure measurements used when modern meters were not available. The 50 Ω resistors are the noise source—the first one at boiling liquid nitrogen temperature (77 K) and the second one at the T_A temperature (see Fig. 9.12, [10]). In this case, $\mathrm{ENR} = (T_A - T_{N_2})/T_A = (290 - 77)/290 = 0.73$, and the noise figure is $F = 0.73/(1 - Y)$, where Y is the ratio of the powers measured in both states at the output of the two-port.[7] This approach is still used for the immediate calibration of some types of radiometers.

9.1.4 Low-Noise Amplifier for 1.3 GHz

The two-stage amplifier with a bandpass filter between the stages and the noise-matched first stage were designed by Ansoft Designer. A design example is present here. Both stages are equipped with pHEMT ATF-55143 $V_{ds} = 2$ V, $I_d = 10$ mA. The LNA was designed on Duroid 5 880 substrate with $\varepsilon_r = 2.2$ and thickness 31 mils (0.79 mm).

7 $Y < 1$ in this case.

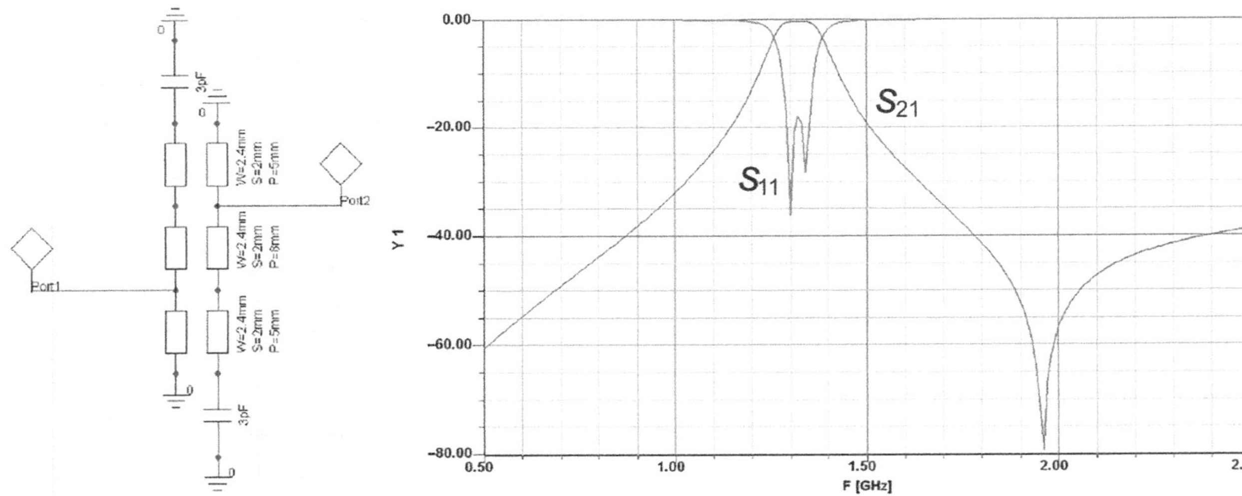

Figure 9.13 Bandpass filter designed from two coupled microstrips tuned by capacitance.

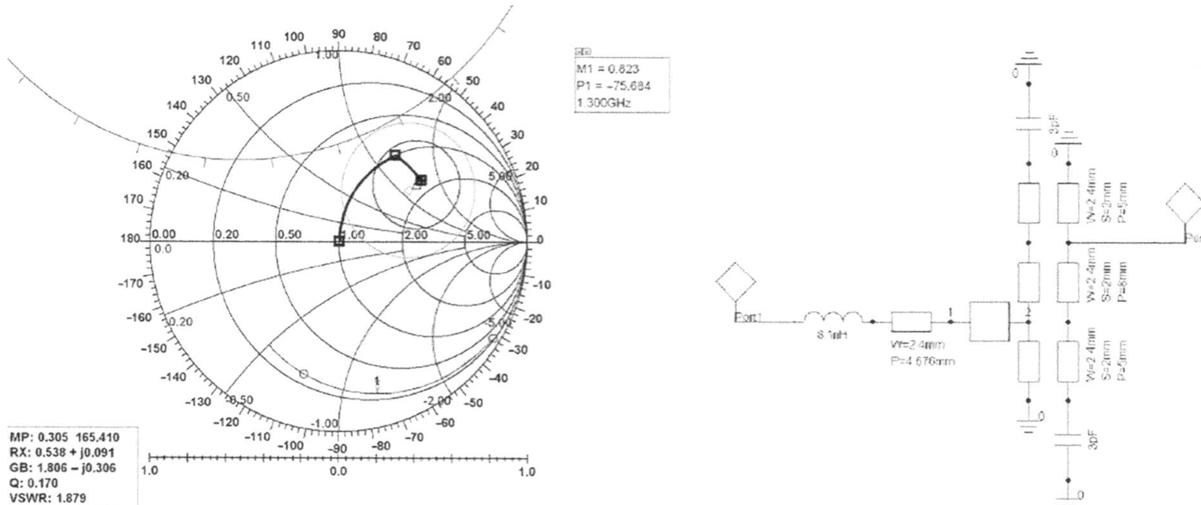

Figure 9.14 Noise matching of the first transistor using an 8 nH inductor and a 50 Ω/10° microstrip. The arc at the top of the Smith chart indicates the stability zone in the generator plane.

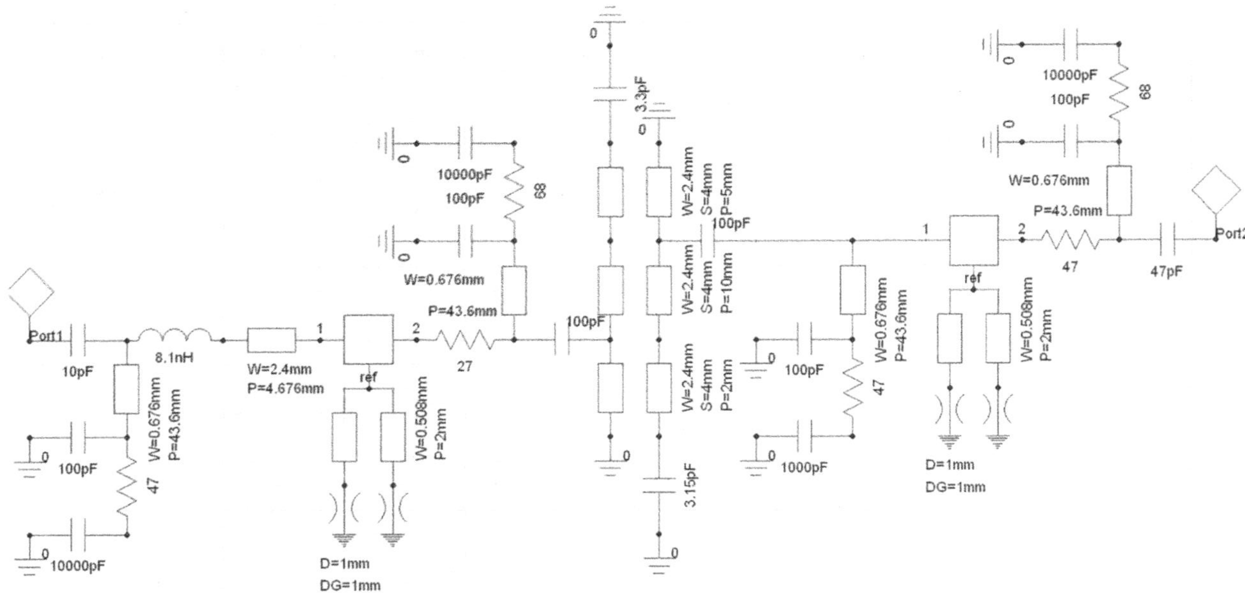

Figure 9.15 Schematic of the two-stage amplifier including power supply circuits (in simulator).

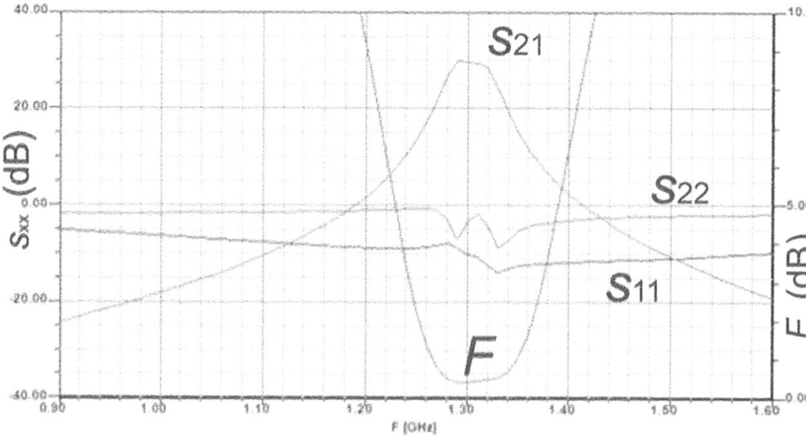

(a)

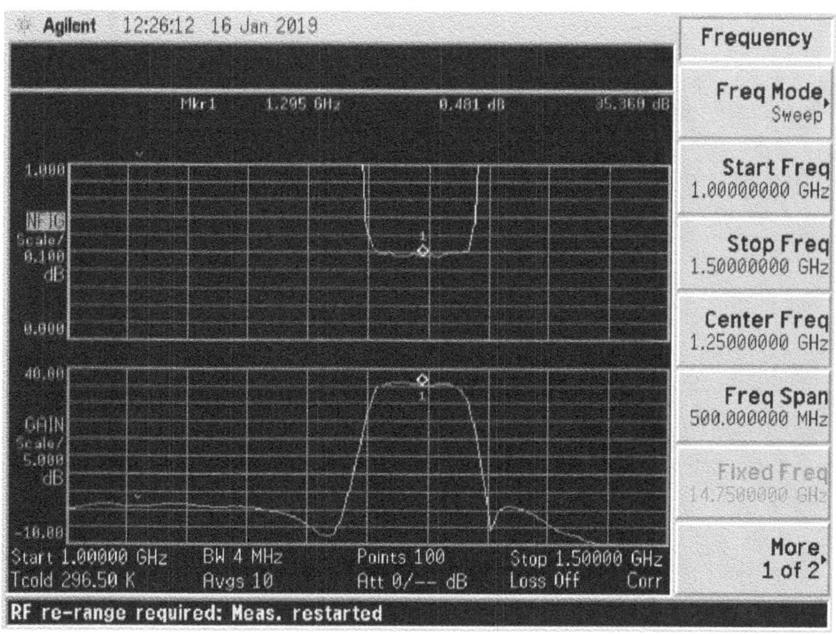

(b)

Figure 9.16 Simulated (a) and measured (b) parameters of LNA for 1 296 MHz.

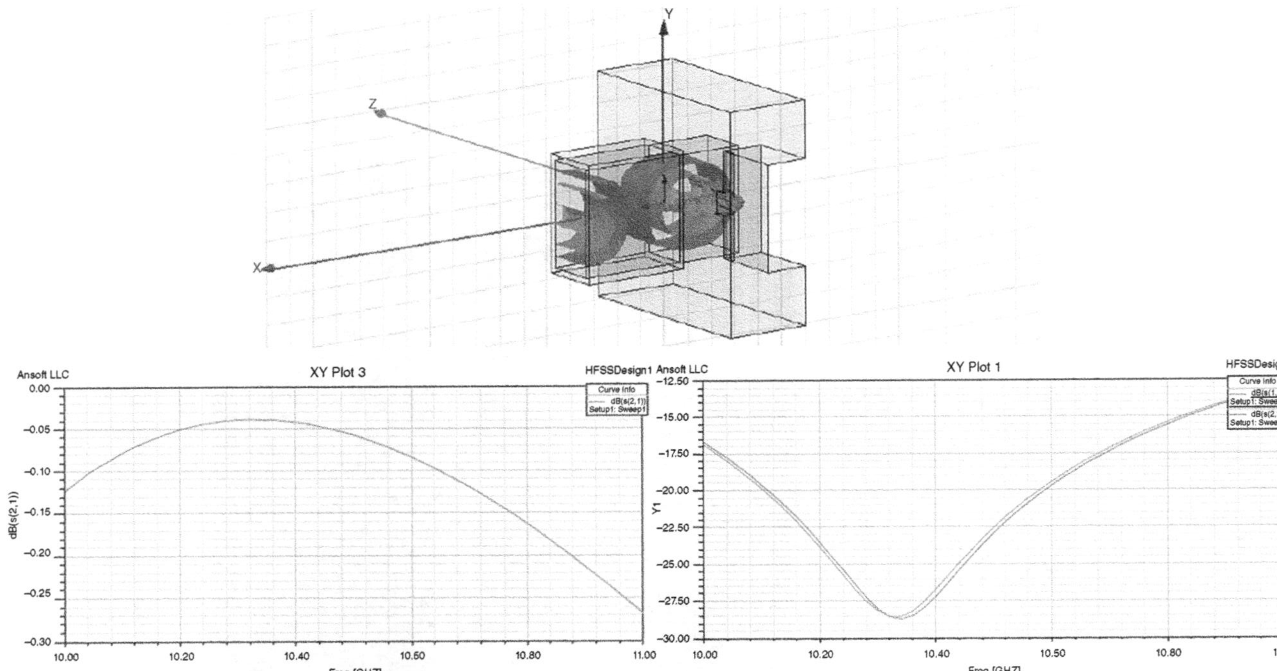

Figure 9.17 Simulated transition from the R120 waveguide to the substrate (top). Power transfer of the junction (bottom left). Reflection coefficient at the input of the waveguide and the output of coupling from the waveguide (bottom right).

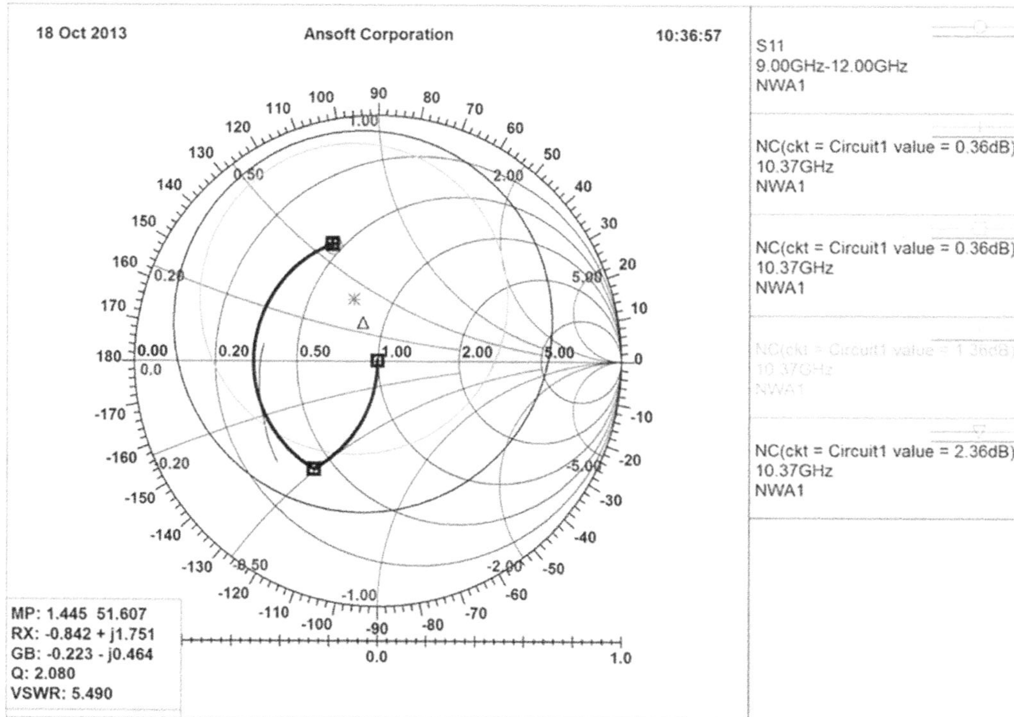

Figure 9.18 Noise matching of the first pHEMT NE3511S02 using a 50 Ω/50° transverse stub and a 50 Ω/64° longitudinal line.

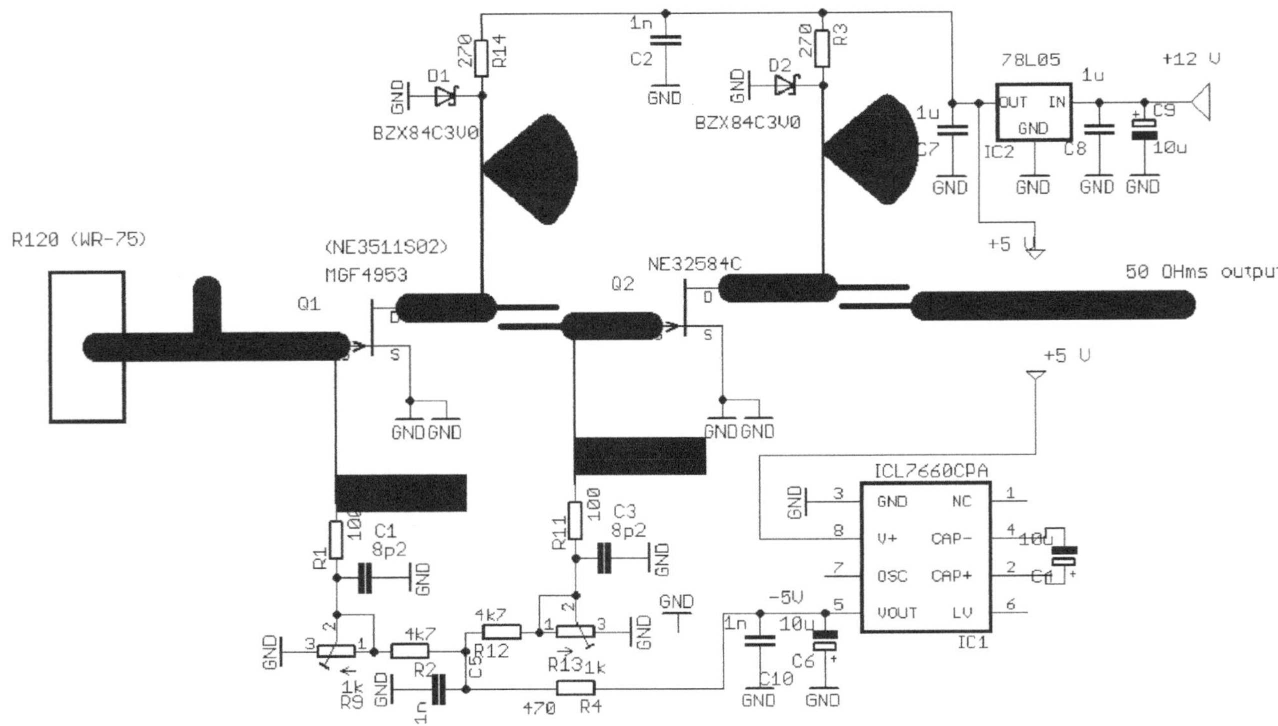

Figure 9.19 Schematic of the LNA for 10.4 GHz.

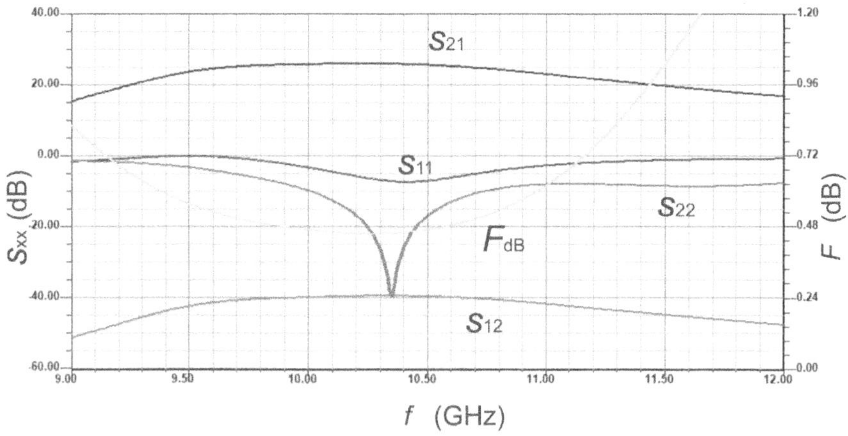

Figure 9.20 Simulated LNA parameters.

A separately designed bandpass filter is shown in Figure 9.13. Figure 9.14 shows the noise matching solution of the first transistor.[8] The schematic of the amplifier, including the supply circuits, is shown in Figure 9.15. Figure 9.16 shows simulation results.

9.1.5 Low-Noise Amplifier for 10 GHz

To achieve high sensitivity, the LNA for 10.4 GHz is designed with a waveguide input. The transition from the waveguide to the substrate with the first transistor was simulated in the ANSYS HFSS. Results of the optimized solution are shown in Figure 9.17. The amplifier consists of two stages. At the input, pHEMT NE3511S02 is used. At the second stage, NE32584C is exploited. Both transistors operate at the point $V_{ds} = 2$ V and $I_d = 10$ mA.

Noise matching is shown in Figure 9.18. Figure 9.19 shows the schematic and Figure 9.20 the simulated parameters of the amplifier by Ansoft Designer. Figure 9.21 shows the design of the amplifier including the waveguide input. Figure 9.22a shows the measured noise figure and gain. A photograph of the amplifier is shown in Figure 9.22b.

8 The arc at the top of the Smith chart in Figure 9.14 (stability zone in the generator plane) defines the range of impedances where the input of the amplifier is not undamped to become unstable.

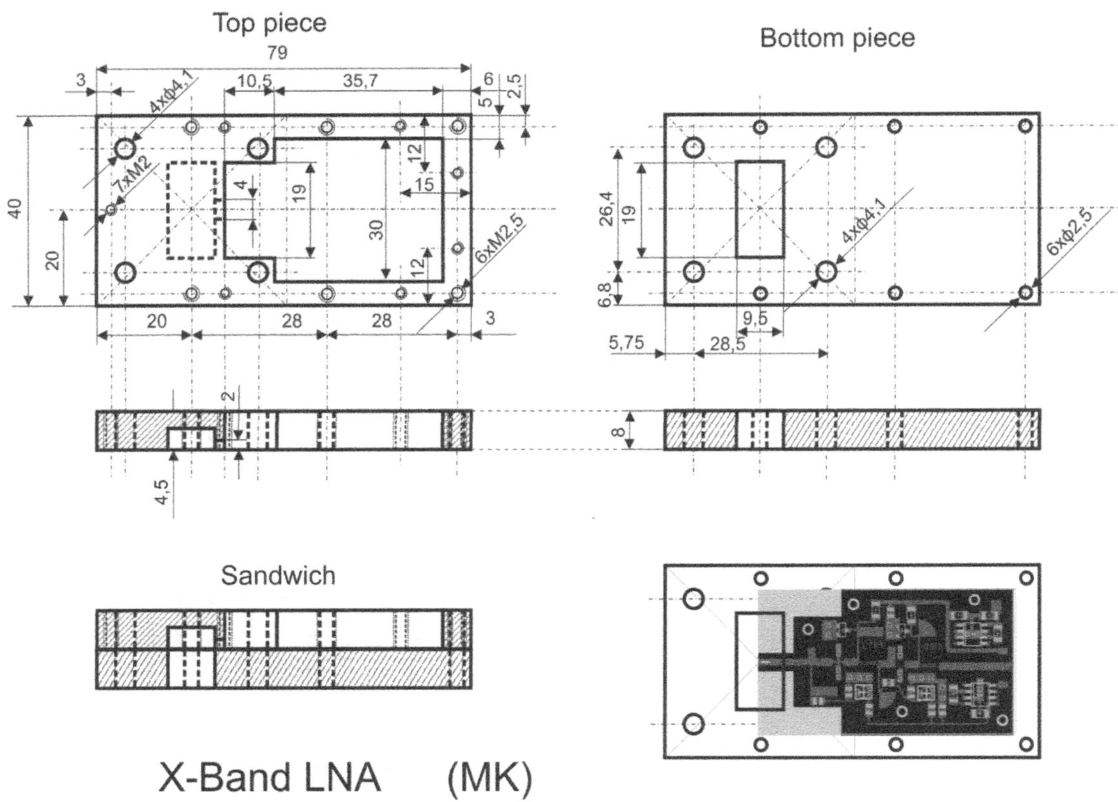

Figure 9.21 LNA design solution for 10.4 GHz.

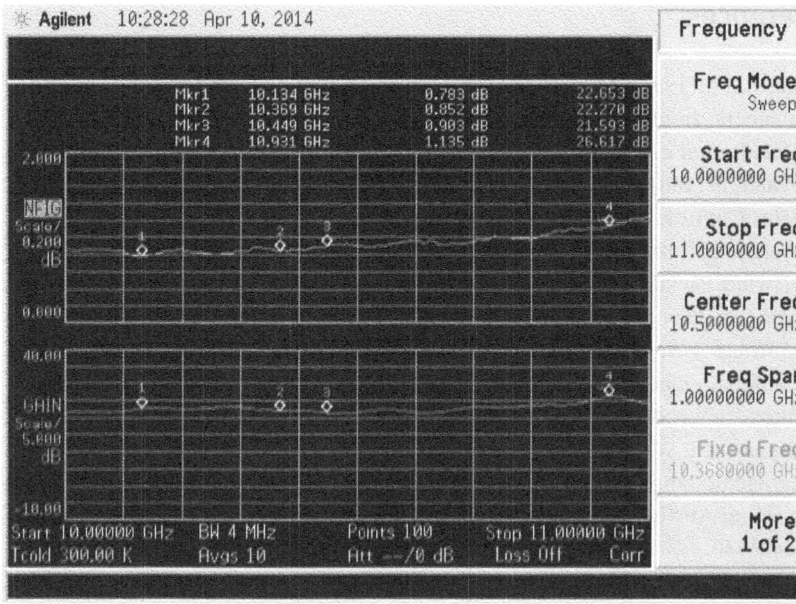

(a)

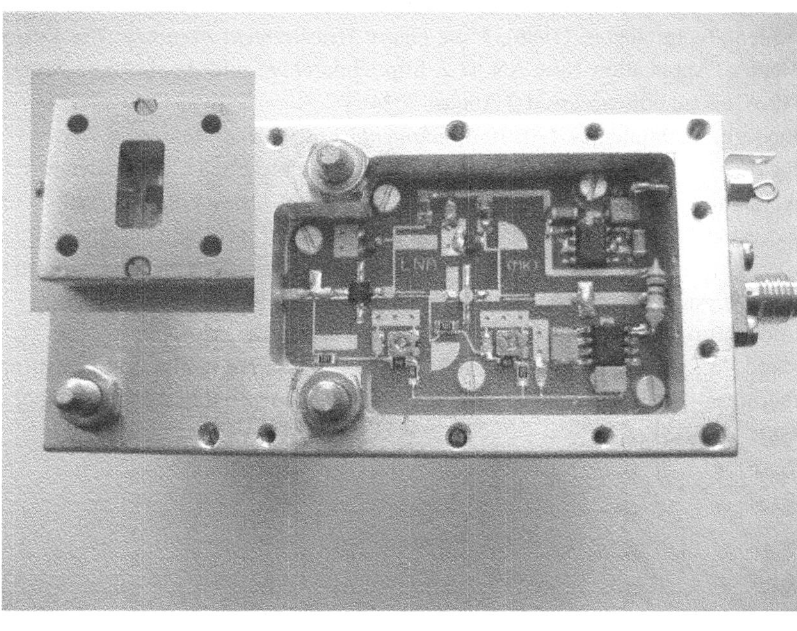
(b)

Figure 9.22 (a) Measured LNA parameters: $F = 0.8$ dB and $A = 22$ dB; (b) LNA design for 10 GHz. *Source: radio.feec.vutbr.cz.*

References

1 Kasal, M. (1998). Modern methods of generation and processing of nuclear magnetic resonance signals (in Czech). *In: Scientific Writings of BUT, Habilitation and Inaugural Writings*. 1: 3–28. Brno, CR: VUTIUM. ISBN 9788021411661.

2 Mini Circuits. (2002). *RF Frequency Mixers*. http://www.minicircuits.com/WebStore/Mixers.html (accessed 7 August 2024).

3 Hewlett Packard. (1998). *Low Cost Mixer for the 10.7 to 12.8 GHz Direct Broadcast Satellite Market*. Application Note 1136. http://www.hp.woodshot.com/hprfhelp/4_downld/lit/diodelit/an1136.pdf (accessed 7 August 2024).

4 California Eastern Laboratories. (2019). *Datasheet of the CE3512K2 Device*. https://www.cel.com/components/low_noise_devices (accessed 7 August 2024).

5 Bryerton, E.W., Morgan, M., and Pospieszalski, M.W. (2013). Ultra-low noise cryogenic amplifiers for radio astronomy. *2013 IEEE Radio and Wireless Symposium*, Austin (TX, USA), 2013: 358–360. https://doi.org/10.1109/RWS.2013.6486740.

6 Low Noise Factory. http://www.lownoisefactory.com/products (accessed 7 August 2024).

7 Agilent Technologies. (1988). *Noise Figure Measurement Accuracy: The Y-Factor Method*. Application Note AN 57-2. http://hparchive.com/Application_Notes/HP-AN-57-2.pdf (accessed 7 August 2024).

8 Keysight Technologies. (2019). *Fundamentals of RF and Microwave Noise Figure Measurements*. Keysight Application Note 5952-8255. https://www.keysight.com/zz/en/assets/7018-06808/application-notes/5952-8255.pdf (accessed 7 August 2024).

9 Keysight Technologies. (2014). *Noise Figure Uncertainty – Calculator*. https://www.keysight.com/us/en/lib/resources/training-materials/noise-figure-uncertainty-calculator-96887.html (accessed 7 August 2024).

10 Kasal, M., Húsek, V., and Halámek, J. (1993). A simple method of measuring the noise figure of the NMR spectrometer. *Radioengineering* 2(2): 5–8. ISBN ISSN 1210-2512.

10

Local Oscillators

Local oscillators are a very important part of each communication system, serving to translate the microwave (radio frequency) signal into the baseband,[1] both in systems based on a heterodyne principle and in systems with sampling [1]. A local oscillator (LO) usually generates a harmonic signal. Basic LO parameters are power level, accuracy of frequency, stability of frequency, and the content of undesired components. Undesired components are of two types: components caused by harmonic distortion of the signal (higher harmonics) and components caused by short-term instability of the signal characterized by phase noise or a quantity called jitter. Higher harmonics usually can be filtered out easily. But the phase noise, which broadens the spectrum of the signal of the local oscillator, cannot be removed, and further processing usually worsens the phase noise level. Phase noise is usually characterized by the power spectral density relative to the carrier level at a given offset from the carrier N_{ph} (dBc/Hz@f_{off}). Multiplying the frequency of the oscillator with the phase noise N_{ph} n times worsens the phase noise level as follows:

$$N_{ph\text{-}n} = N_{ph} + 20 \log n. \tag{10.1}$$

This applies to frequency synthesizers, where N_{ph} is a phase noise of the fundamental oscillator the synthesizer relies on. Next, n is the ratio between the resulting frequency and frequency of the reference oscillator.

In today's communication systems, the local oscillator is more and more frequently implemented by a frequency synthesizer. The technology of integrated circuits has reached such a quality that complex synthesizers can be implemented in minimal dimensions and with minimal power consumption even at GHz frequencies. Moreover, mass production of some types of integrated circuits for frequency

1 Often, several transposition oscillators are used, and additional intermediate frequency bands are created.

Reception of Weak Radio Signals from Space, First Edition. Miroslav Kasal.

synthesizers (e.g. for consumer electronics and mobile phones) made these circuits very cheap. As a consequence, a radio receiver with a simple frequency synthesizer is cheaper than a receiver with a classical tunable local oscillator. And some modern communication systems cannot be implemented without frequency synthesis, e.g. systems with frequency hopping or systems operating with extremely low signal power requiring coherent detection methods. In the following discussion, two methods of frequency synthesis are presented, which currently are the most frequently used (including an analysis of spectral properties of generated signals). The final overview presents some of our applications of frequency synthesis in communication systems.

10.1 Classification of Frequency Synthesizers

10.1.1 Coherent and Incoherent Synthesizers

The phase relation between the generated signals and the reference signal (the degree of coherence) is probably the most important property from the viewpoint of generated signals. But evaluation of the degree of coherence is quite complicated. Whether the phase relation between the signals is defined or not is given by the basic concept of the synthesizer. From this viewpoint, synthesizers are divided into coherent and incoherent ones.

10.1.2 Synthesizers with Direct and Indirect Synthesis

Next, frequency synthesizers with direct synthesis are distinguished. In their case, the resulting signal is created by basic arithmetic operations (addition, subtraction, multiplication, division) at the frequency of reference signal(s). Synthesizers with direct digital synthesis (DDS) are very important. The time response of a generated signal is read from the memory at a corresponding speed. In digital systems, generated signals can be used in binary form directly (e.g. for complex multipliers) or can be converted to analog signals using a D/A converter and a subsequent low-pass filter. Operation of frequency synthesizers with indirect synthesis is based on the principle of a feedback system, most often a phase-locked loop (PLL) (see Fig. 10.1).

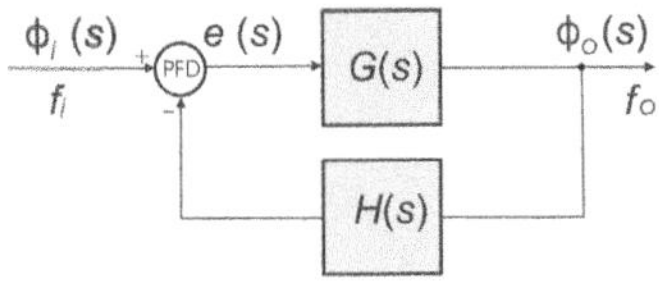

Figure 10.1 A model of control system with negative feedback as a basis of frequency synthesizers with indirect synthesis.

10.2 Direct Digital Frequency Synthesizers - DDFS

10.2.1 Basic Description

The technology of direct digital frequency synthesis has progressed tremendously in recent years. The principle is well known (see Fig. 10.2). Using the phase accumulator, ROM or RAM memory addresses are generated to store the generated waveform table. A digital signal is converted to an analog one by a D/A converter and a subsequent anti-aliasing low-pass filter. The synthesizer has two digital inputs—the clock signal of the frequency f_C and the frequency setting word (FSW) of the phase accumulator. The output frequency is given by the relation

$$f_o = \frac{\text{FSW}}{2^N} f_C, \quad \text{FSW} \in \{1,\ 2^{N-1}\}, \tag{10.2}$$

where N is the number of bits of the control word FSW of the phase accumulator. The frequency of the output signal is in principle limited, $f_C/2^N < f_o < f_C/2$, with resolution $f_C/2^N$. At each clock cycle, the output value of the phase accumulator is increased by the step ΔPhase.

The output value of the phase accumulator in binary form determines the phase of the generated harmonic signal. For all zeros, the phase is 0 rad, and for all ones, the phase is 2π rad, i.e. $\Delta\text{Phase} = 2\pi/2^N$. In the next block, the sine function is generated taking the M most significant bits from the output of the phase accumulator and creating an (M–2)-bit word, which corresponds to the instantaneous amplitude of the output signal. The value

$$D = M - 2 \tag{10.3}$$

is given by the resolution of the following converter (or the digital multiplier). Trimming of the $(N - M)$ least significant bits is necessary because D/A converters usually have lower resolution. The DDS output signal does not have to be harmonic. On the contrary, the signal can be of any periodic waveform. To convert the phase to a functional value, a memory or a processor with an appropriate algorithm can be used. At the output of this block, the generated signal is in digital

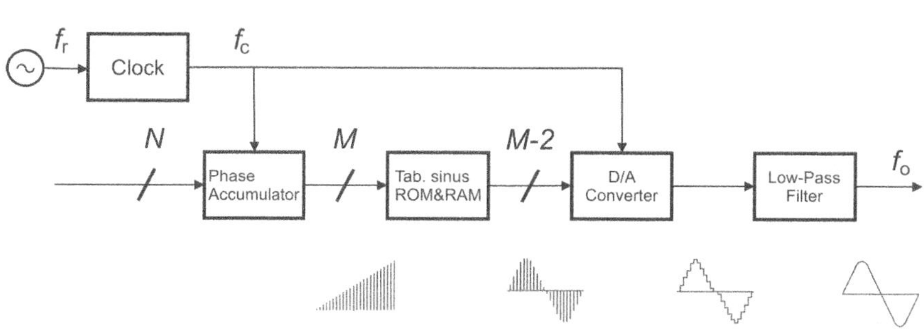

Figure 10.2 Block diagram of a direct digital synthesizer.

form. In modern systems with DSP, the signal can be directly used for several applications as shown below.

10.2.2 Spectral Properties of DDFS

When generating an analog signal, the D/A converter is usually a limiting factor in current DDFS systems. Today, 16-bit monolithic DAC with sampling rates up to 600 Msps are available. As shown later, the given speed and resolution parameters of the converter are not sufficient to assess the quality of the generated signal. For the SNR ratio (dB) corresponding to the quantization noise (± 0.5 LSB), we can theoretically write

$$\text{SNR} = 1.76 + 6.02D, \tag{10.4}$$

where D is the number of DAC bits. In practice, the 6 dB/bit offset is expected.

Quantization noise can be expressed as a relative power spectral density (dBc/Hz) in the frequency range $(0, f_C/2)$

$$N_q = -[1.75 + 6.02D + 10\log(f_C/2)]. \tag{10.5}$$

Quantization noise is expressed in the same units as the phase noise level. Both types of noise can be therefore easily compared. However, these values can usually be achieved only at very low frequencies, i.e. under the condition $f_o \ll f_C/2$. In the output signal spectrum, discrete components can be usually identified. Therefore, it's better to describe the properties of the generated signal with the spurious free dynamic range (SFDR, see Fig. 10.3). The selection of a D/A converter itself is not simple. DC offset, gain error (for a full dynamic range), differential nonlinearity (DNL), and integral nonlinearity (INL) are usual parameters describing the transfer function of D/A converters. For a 3-bit D/A converter, these parameters are

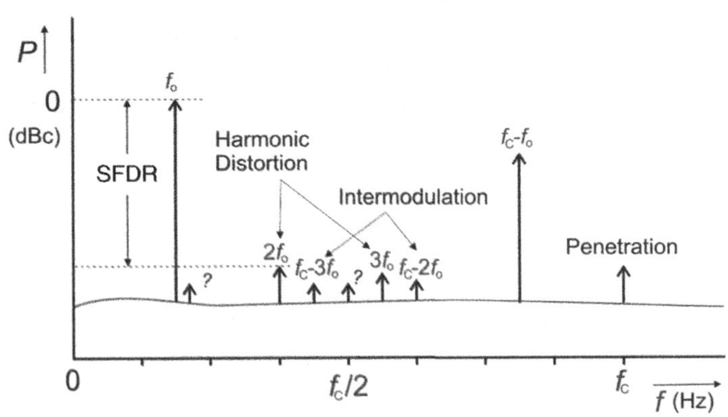

Figure 10.3 Typical spectrum of a direct digital synthesizer.

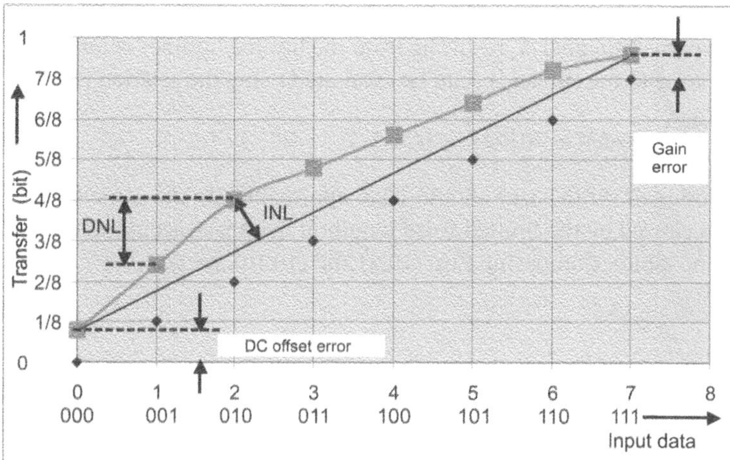

Figure 10.4 Transfer function of a 3-bit D/A converter to illustrate DNL and INL.

shown in Figure 10.4 (with simplification). The influence of these parameters on the spectral properties of the generated signals is not explicit. Integral nonlinearity INL of 0.5 LSB over the entire dynamic range of the transducer causes a higher SFDR decrease than INL of 1 LSB occurring at a single value (e.g. in practice with Qualcomm DDFS synthesizer, better results have been obtained using 10-bit Sony D/A converter than 12-bit D/A converter by TRW Company).

In [2, 3], V. Kroupa derived a relation for the power spectral density (dBc/Hz), which is similar to Eq. (10.5) but considers the limited word length in memory

$$N_\mathrm{d} = -[7.8 + 6D + 10\log(f_\mathrm{C}/2) \pm 10]. \tag{10.6}$$

The amplitude of the odd harmonic components can be estimated ($i = 3, 5, \ldots$)

$$A_i = 2\frac{2^{-D}}{\pi}\frac{1}{i}. \tag{10.7}$$

The first to third harmonics ratio is (in logarithmic scale)

$$20\log\left(\frac{A_3}{A_1}\right) \approx -6D - 13.5. \tag{10.8}$$

Next to D/A converters, even other phenomena play a role in the generation of digital signals. Limitation of the word length at the output of the phase accumulator is important. Contrary to the limitation of the word length in memory and ADC, this limitation introduces phase noise. As already stated, the M most significant bits are taken from N bits of the phase accumulator. This means that the B least significant bits are trimmed according to the relation

$$M = N - B. \tag{10.9}$$

The consequence is the phase modulation of a sine (harmonic) wave. The level of the strongest parasitic signal A_s resulting from the limited number of bits of the accumulator related to the carrier A_c can be estimated using the relation [3]

$$20\log\left(\frac{A_s}{A_c}\right) \approx -6M - 20\log r, \quad (r = 1, 2, \ldots, 2^{B-1}). \tag{10.10}$$

A properly designed synthesizer should have approximately the same level of parasitic signals caused by the limited word length of the phase accumulator and the memory (sine table). Comparing Eqs. (10.8) and (10.10), we get

$$-6M \approx -6D - 13.5, \tag{10.11}$$

and from here

$$M \geq D + 2. \tag{10.12}$$

This is the derivation of Eq. (10.3), which was given at the beginning of this chapter.

When verifying parameters of the practical implementation of DDS, a measured SNR is considered. The value D calculated from Eq. (10.4) is denoted as D_{eff}, or the effective number of bits, while $D_{\text{eff}} \leq D$.

DDFS applications with a digital output were already mentioned. Several modern communication systems use a Hilbert transformer (HT) to transform signals from or to the baseband. Amplitude symmetry and quadrature accuracy of components I and Q determine the suppression of mirror products. An amplitude asymmetry of 1% and phase inaccuracy of orthogonality of 1° can be easily shown to limit the suppression of mirror frequency rejection to ~40 dB. In conventional analog systems, these values are just the achievable limit at relatively low frequencies. DDFS are directly intended for the generation of quadrature signals. A single-phase accumulator is sufficient to synchronously control two memories with a corresponding shift. If complex digital multipliers are used instead of analog ones, D/A converters are no longer needed. A very accurate HT can be implemented fully digitally with mirror frequency suppression better than 80 dB at frequencies of tens of megahertz. Figure 10.5 shows the block diagram of such a system, which uses HT in the digital tuner and a digital Costas loop in the detector. The system can form the core of software-defined radio (SDR). Each DDFS block is denoted as a numerically controlled oscillator (NCO).

Our laboratory achieved reasonable experimental results. The telemetry and command system was designed to generate signals precisely defined in terms of frequency and even phase (if possible). Local oscillators of all microwave converters have to be therefore controlled by a single frequency standard. From several options, GPS was selected. The converter for the 2.4 GHz band is implemented then according to Figure 10.6.

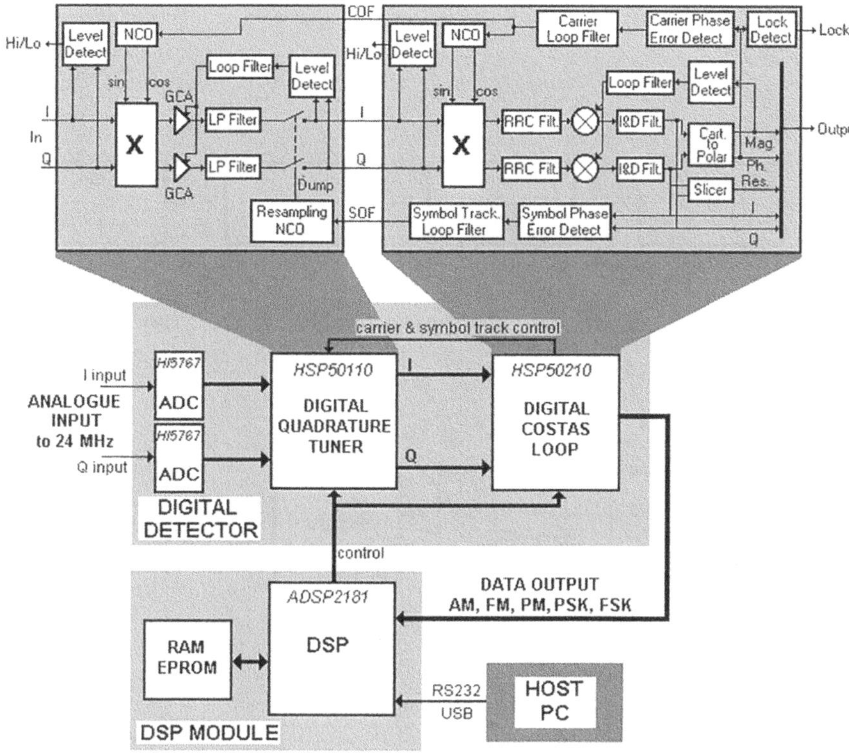

Figure 10.5 Block diagram of a digital receiver.

For this system, a local oscillator was designed, synchronized with the GPS signal and achieving an appropriate accuracy of frequency setting and appropriate spectral purity. A solution is based on the AD9857 chip by Analog Devices, which contains a 14-bit CMOS digital quadrature up-converter. From the chip, only PLL and DDFS with a 14-bit D/A converter were used to generate a synchronized 200 MHz clock (see Fig. 10.7). Since the status word of the phase accumulator (frequency setting word, FSW) is 32 bits long, the resulting frequency can be set with an accuracy better than 0.05 Hz. The spectral purity of the output signal was examined in detail.

According to Eq. (10.5), the spectral density of the quantization noise (14 bits) should be −166 dBc/Hz. Considering the thermal noise (kT_0) being −174 dBm/Hz at a temperature of 300 K, both components are approximately comparable at a level of the generated signal around 0 dBm. According to Eq. (10.10) (for $r = 1$), the level of discrete parasitic components should be −84 dBc. The manufacturer states SFDR >80 dB for 65 MHz ± 100 kHz around carrier. The value was confirmed by measuring under similar conditions (a narrow-band filter

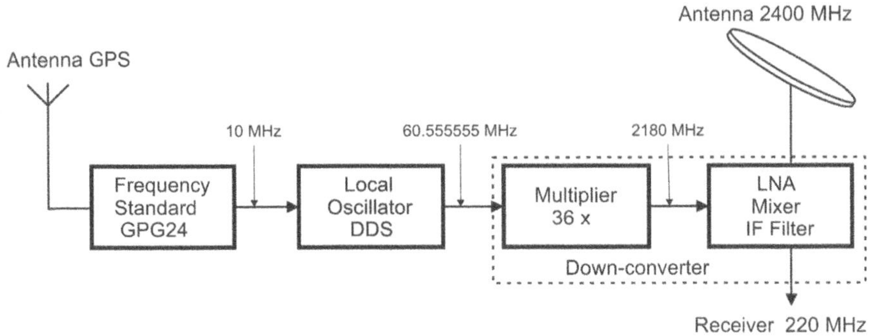

Figure 10.6 Receiving system.

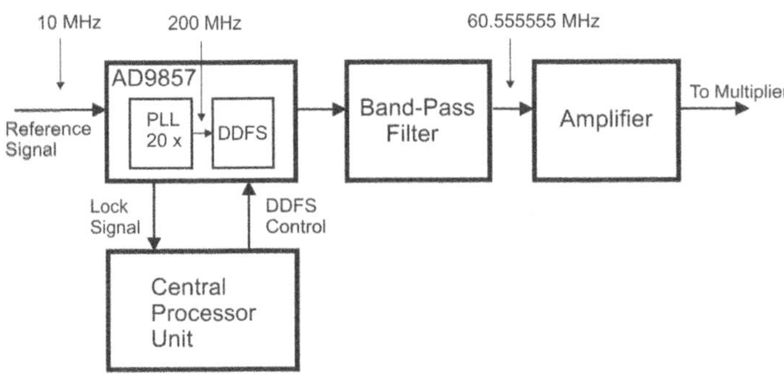

Figure 10.7 DDFS as a local oscillator synchronized by an external signal.

at the output) (see Fig. 10.8). The measured phase noise values are shown in Figure 10.9.[2]

10.3 Phase-Locked Loop Synthesizers

10.3.1 Time and Frequency Domain Description of a Phase-Locked Loop

Synthesizers at higher frequencies (up to the low tens of GHz) can be implemented by a synthesizer with a PLL [4, 5]. For the basic description, the arrangement from Figure 10.10 is used. The input and output voltages are as follows

$$u_i(t) = U_i \sin[\omega_i t + \phi_i(t)] \quad \text{and} \quad u_o(t) = U_o \sin[\omega_o t + \phi_o(t)]. \tag{10.13}$$

2 High-frequency direct digital synthesizers require a high reference frequency, which can be usually obtained using a phase-locked loop. From the viewpoint of phase noise, this influences the characteristics of the synthesizer.

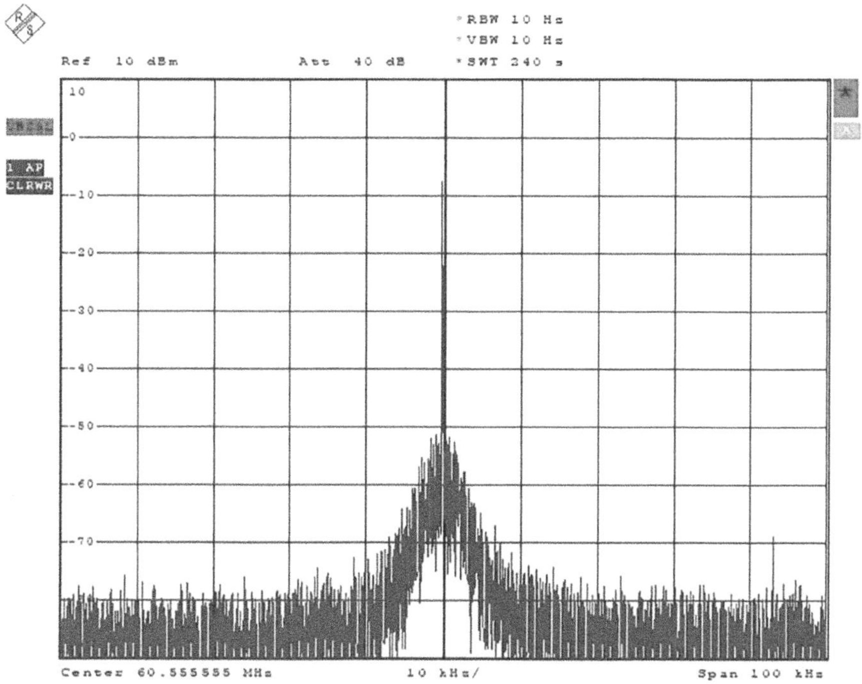

Figure 10.8 DDFS spectrum with an AD9857 circuit.

The phase $\phi_i(t)$ is chosen as a reference and set to $\phi_i(t) = 0$. At the output of the phase detector, which is a multiplier in the simplest case, we get

$$u_d(t) = K_1 U_i U_o (\sin \omega_i t)[\sin(\omega_o t + \phi_o(t))]. \tag{10.14}$$

Now, two cases are distinguished:

1) The loop is not locked $\omega_i \neq \omega_o$
   Equation (10.14) yields

$$u_d(t) = \frac{1}{2} K_1 U_i U_o \{\cos[(\omega_i - \omega_o)t - \phi_o(t)] - \cos[(\omega_i + \omega_o)t + \phi_o(t)]\}, \tag{10.15}$$

   and after filtration (and amplification)

$$u_v(t) = K_2 U_i U_o \{\cos[(\omega_i - \omega_o)t - \phi_o(t)]\}. \tag{10.16}$$

   At $u_v(t) = 0$, the voltage-controlled oscillator (VCO) oscillates with frequency ω_o'. If the harmonic voltage, created as a beat between both the frequencies, impacts the VCO, the frequency ω_o' is shifted by $\pm\Delta\omega(t)$. Assuming the

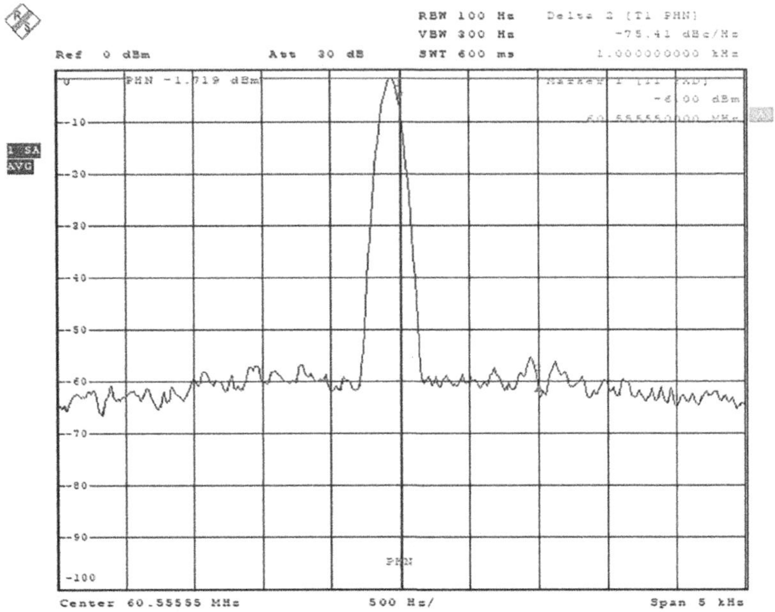

(a)

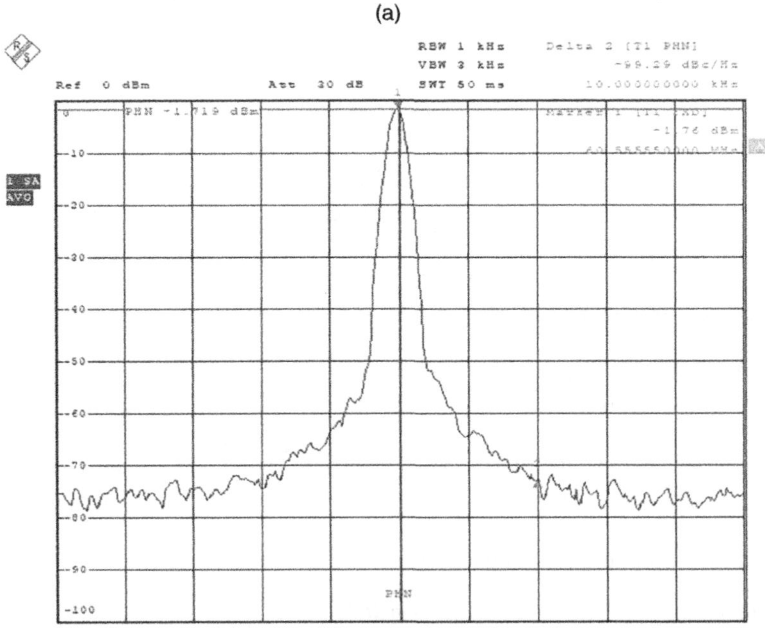

(b)

Figure 10.9 DDFS-LO phase noise measurement: (a) −75 dBc/Hz @ 1 kHz; (b) −99 dBc/Hz @ 10 kHz.

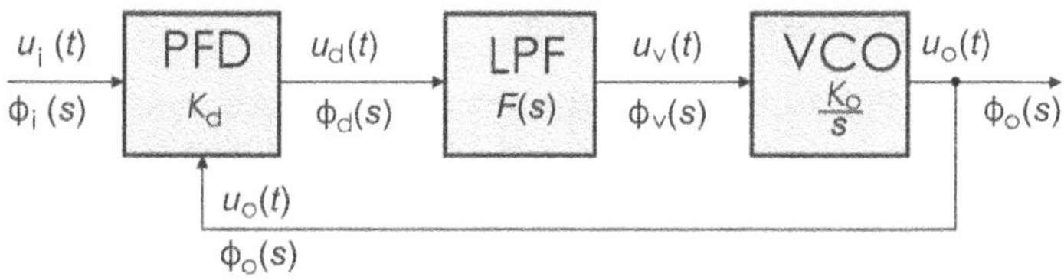

Figure 10.10 Phase-locked loop (VCO is a voltage controlled oscillator, PFD a phase-frequency detector).

amplitude $K_2 U_i U_o$ is high enough and no limitations appear, the system settles down to hold

$$\omega_i = \omega_o = \omega_o' \pm \Delta\omega, \tag{10.17}$$

and the PLL is locked. The condition of sufficient beat amplitude (without limitation or saturation) means that the quantity $\Delta\omega$ is finite and defines the capture band of the PLL.

2) The loop is locked $\omega_i = \omega_o$
Equation (10.14) can be then modified as follows

$$u_d(t) = \frac{1}{2} K_1 U_i U_o \{\cos\phi_o(t) - \cos[2\omega_o t + \phi_o(t)]\}, \tag{10.18}$$

and after filtration (and amplification),

$$u_v(t) = K_2 U_i U_o \cos\phi_o(t). \tag{10.19}$$

If $\omega_i = \omega_o = \omega_o'$ (i.e. frequency of a free-running VCO is tuned to the expected frequency), then

$$u_v(t) = 0 \quad \text{and thus} \quad \phi_o(t) = \pm\frac{\pi}{2}.$$

In other words, the voltages u_i and u_o are mutually phase-shifted for 90° in steady (locked) state. If now ω_i differs from ω_o' slightly, the phase shifts for $\Delta\phi(t)$, i.e.

$$\phi_o(t) = \pm\frac{\pi}{2} + \Delta\phi(t), \tag{10.20}$$

and Equation (10.19) is of the form

$$u_v(t) = K_2 U_i U_o \sin\Delta\phi(t). \tag{10.21}$$

Since sine is an odd function, the direction of tuning the VCO is indicated to keep the loop locked. Equation (10.20) implies that $-\pi/2 < \Delta\phi(t) < \pi/2$, which corresponds to $\phi_0(t) \in (0; \pi)$.

10.3.2 Order and Type of Phase-Locked Loop

To find a response of the PLL to the input disturbance, a Laplace transform is used (assuming the system is linear). For the gain of the closed-loop system in Figure 10.1, the following well-known relation applies

$$I(s) = \frac{G(s)}{1 + G(s)H(s)}. \tag{10.22}$$

Here, $G(s)$ is the gain in the direct branch, and $G(s)H(s)$ is the gain of the open loop. For harmonic signals, $s = \mathrm{j}\omega$.

For separate signals of phase lock, we can write

$$\Phi_\mathrm{d}(s) = [\Phi_\mathrm{i}(s) - \Phi_\mathrm{o}(s)]K_\mathrm{d}, \tag{10.23}$$

$$\Phi_\mathrm{v}(s) = \Phi_\mathrm{d}(s)F(s), \tag{10.24}$$

$$\Phi_\mathrm{o}(s) = \Phi_\mathrm{v}(s)\frac{K_\mathrm{o}}{s}, \tag{10.25}$$

and for the transfer function of the PLL according to Eq. (10.22),

$$I(s) = \frac{\Phi_\mathrm{o}}{\Phi_\mathrm{i}} = \frac{K_\mathrm{d}K_\mathrm{o}F(s)}{s + K_\mathrm{d}K_\mathrm{o}F(s)}. \tag{10.26}$$

Here, K_d(V/rad) is the conversion gain of the phase detector, and K_o (rad sec^{-1} V^{-1}) is the conversion gain of the voltage-controlled oscillator.

The degree of the polynomial in the denominator of Eq. (10.26) defines the order of the PLL and affects its behavior at higher frequencies of sidebands (from the carrier). Close to the carrier, i.e. when examining steady values of errors of the PLL, the influence of the type of the phase-locked look prevails. The loop type is determined by the number of poles of the transfer function of the open loop $G(s)H(s)$, i.e. by the number of integrators in the loop.

10.3.3 Phase-Locked Loop Synthesizers

A simple frequency synthesizer with the PLL is created by inserting a frequency divider with dividing ratio N into the feedback branch. Even the reference signal is usually divided by the ratio R to obtain a sufficiently fine step of the synthesizer (see Fig. 10.11). The relation for the output frequency is then

$$\frac{f_\mathrm{o}}{N} = \frac{f_\mathrm{r}}{R} \quad \text{and hence} \quad f_\mathrm{o} = \frac{N}{R}f_\mathrm{r}. \tag{10.27}$$

The expression for the gain of the open loop with the frequency divider is in the form

$$G(s)H(s) = \frac{K_\mathrm{d}K_\mathrm{o}F(s)}{sN}, \tag{10.28}$$

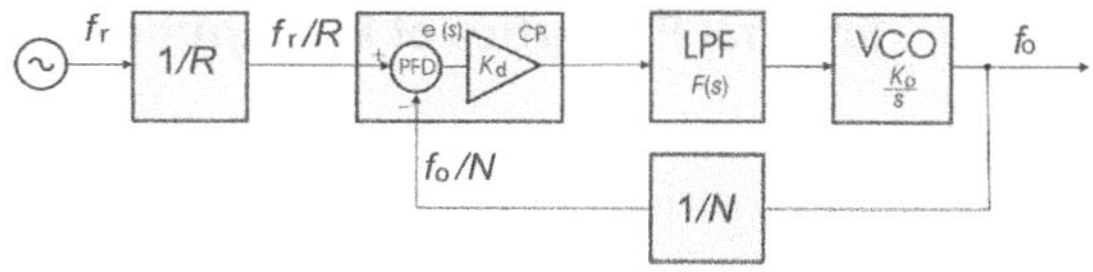

Figure 10.11 PLL frequency synthesizer with frequency dividers.

and Equation (10.26) for PLL transmission becomes

$$I(s) = \frac{NK_dK_oF(s)}{sN + K_dK_oF(s)}. \tag{10.29}$$

10.3.4 Phase-Locked Loop Stability

The stability of the PLL can be investigated using the usual criteria (Hurwitz, Nyquist, etc.). The stability can be easily determined using the Bode plot of module and phase of the gain of the open loop (see Fig. 10.12). The feedback system oscillates if the denominator of the transfer function Eq. (10.22) becomes zero, and thus $G(s)H(s) = -1$. The angular frequency ω_p is the point where the gain of the open loop is unity; it is called the bandwidth of the loop. Phase reserve ψ_m is the phase difference at unity gain and $-180°$. Having the loop stable, the point of unity gain has to be lower in frequency than the frequency at which the phase characteristic reaches the value $-180°$. To ensure stability, the phase characteristics should have a maximum precisely at unity gain so that the phase reserve is maximal (see Fig. 10.12). PLLs of the first and second order are absolutely stable. In practical design and construction, extra poles can be created by various parasitic couplings.

10.3.5 Classification of Phase-Locked Loops

PLL and DPLL are distinguished depending on whether the detector is analog or digital. If the whole PLL is implemented by digital circuits, it is called ADPLL. Such a PLL can be implemented software using DSP technology, and is referred to as SPLL.

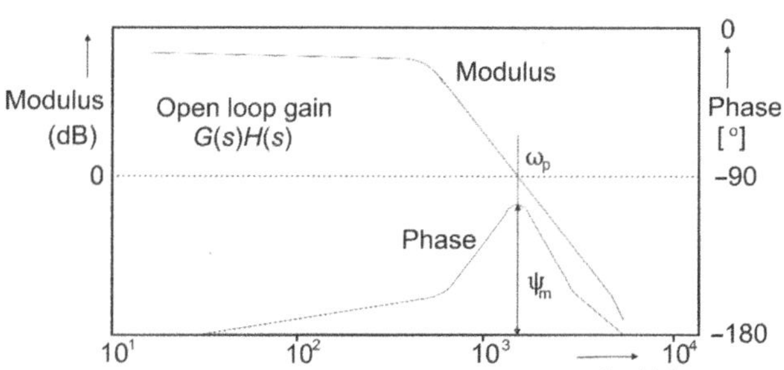

Figure 10.12 Bode plot for determining a phase reserve as a basic criterion of stability.

10.3.6 Spectral Properties of PLL Synthesizers

Unfortunately, the voltage $u_v(t)$, which is delivered to the VCO, can contain noise components of various origins, which causes phase and amplitude modulation of the harmonic signal. However, these parasitic modulations are aimed to be eliminated for a given type of radio communication. In case of PLL synthesizers, two types of parasitic signals are distinguished: the phase noise and discrete parasitic components (see Fig. 10.13). In a well-designed PLL synthesizer, discrete components of the highest level are caused by leaking the signal with reference frequency ΔF.

All parts of the synthesizer contribute (to some degree) to the generation of the phase noise. The level of the phase noise is usually expressed as a power of one sideband per 1 Hz of bandwidth, i.e. power density of the phase noise $N_\nu(f)$. Since the power of the carrier P_c is several orders of magnitude higher than the power of sidebands, the total power of the investigated signal can be assumed to be the power of the carrier and $k(f) = -10\log[P_c/N_\nu(f)]$ (dBc/Hz). Spectral density defined that way is denoted by SSB and is measured with a spectrum analyzer. Since the maximum resolution of spectrum analyzers tends to be higher than 10 Hz, the value at 1 Hz needs to be calculated. The noise bandwidth is about 1.2 times the nominal resolution. A typical spectrum of a PLL synthesizer is shown in Figure 10.14. The PLL creates sidebands around the carrier that have a small overshoot around frequencies $\omega_o \pm \omega_p$ (3–4 dB at 45° phase reserve). The dependence $N_\nu(f)$ decreases at a rate of 20 dB/dec.

The consequence of parasitic components of the spectrum of the generated signal is multivalent. These components can cause interference in adjacent channels (if they are a part of the transmitter) or interference by adjacent channels (if they are a part of the receiver). From the viewpoint of the transmission channel, the phase noise is especially dangerous (discrete components of a higher level are outside the channel) because the dynamics is reduced. The total noise power in a channel Δf wide, caused by the phase noise, is

$$P_{ph} = 2\int_0^{\Delta f} N_\nu(f)\,df. \tag{10.30}$$

If the phase noise level is $k(f) = -80$ dBc/Hz in carrier proximity wider than ±10 kHz, then the dynamics is limited to the value $-(-80+3) - 10\log(10^4) = 37$ dB for a channel $\Delta f = 10$ kHz wide.

An effective phase error E_{rms} caused by phase noise is very important for the transmission of digital signals. The influence of the error is obvious from the constellation diagram for the QPSK signal (see Fig. 10.15). Using introduced symbols,

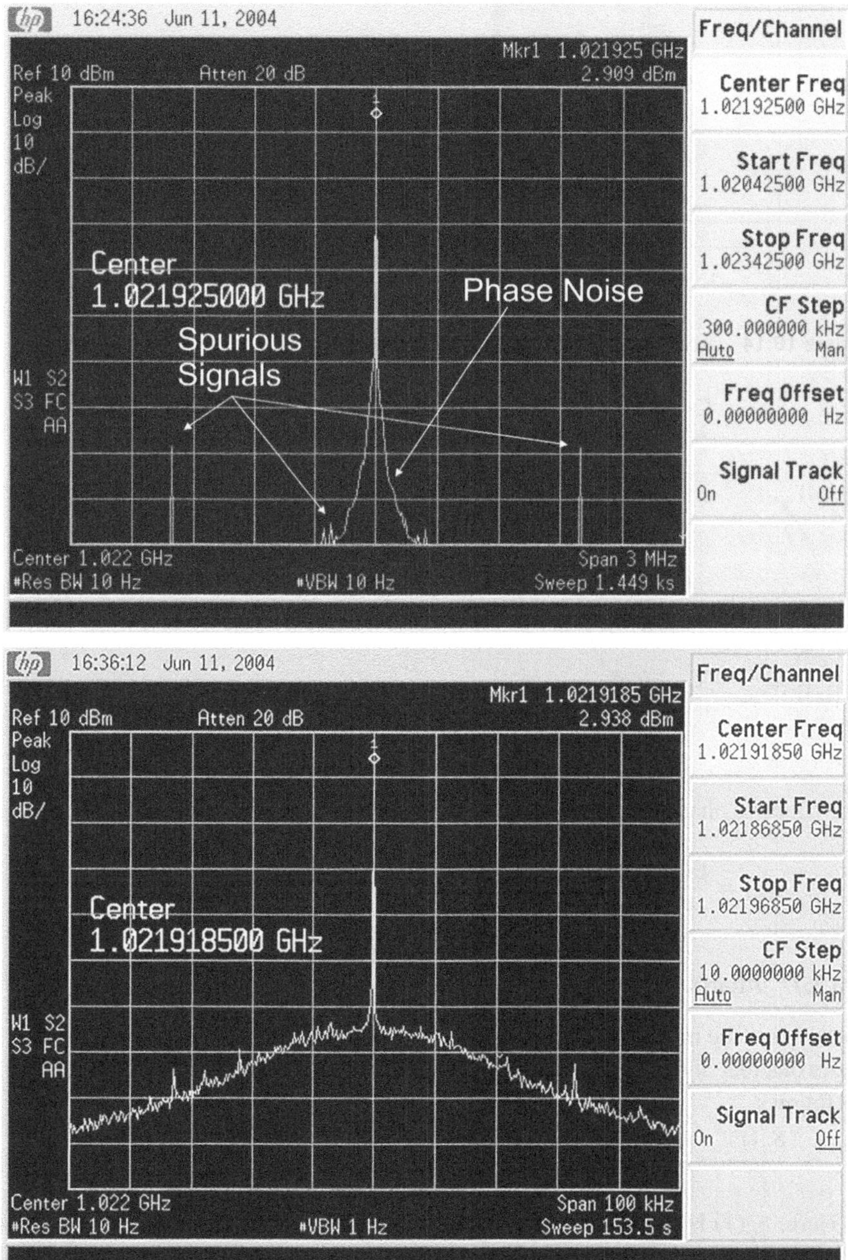

Figure 10.13 Output signal of the PLL synthesizer with frequency 1 GHz (top). Discrete components are generated by the modulation of a leaking signal with reference frequency 1 MHz. The bottom diagram shows the phase noise in the range ±50 kHz from the carrier.

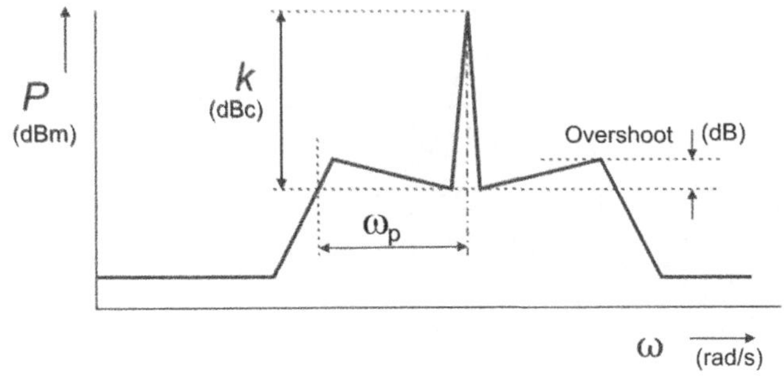

Figure 10.14 A typical spectrum of a signal generated by the phase-locked loop.

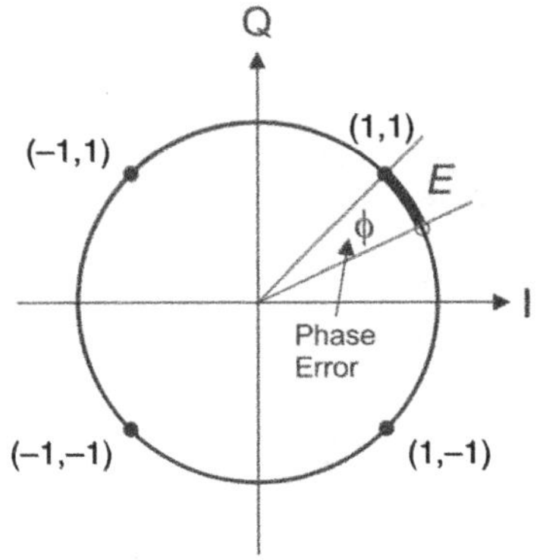

Figure 10.15 The influence of the effective phase error visible in the constellation diagram of the QPSK signal.

the effective phase error is [5]

$$E_{\text{rms}} = \frac{180}{\pi}\sqrt{2\int_a^b N_\nu(f)\text{d}f}. \tag{10.31}$$

10.3.7 Allan's Variance

Let us note the macroscopic noise properties of the oscillator. The noise characteristics of a real high-quality oscillator can be partially linearized ([6], see Fig. 10.16) as follows

$$\frac{S_\varphi(f)}{f_0^2} = \frac{h_{-2}}{f^4} + \frac{h_{-1}}{f^3} + \frac{h_{-0}}{f^2} + \frac{h_1}{f} + h_2. \tag{10.32}$$

Here, $S_\varphi(f)$ is the power spectral density of the output phase. Normalization $S_\varphi(f)/f_0^2$ is introduced to exclude the influence of multiplication and division of the oscillator frequency. Approximately, $S_\varphi \approx 2N_\nu(f)$.

Figures 10.13 and 10.16 show that measuring the spectral density of the phase noise is more difficult closer to the carrier. Therefore, the time domain measurements are used, using the so-called Allan's variances. A harmonic signal

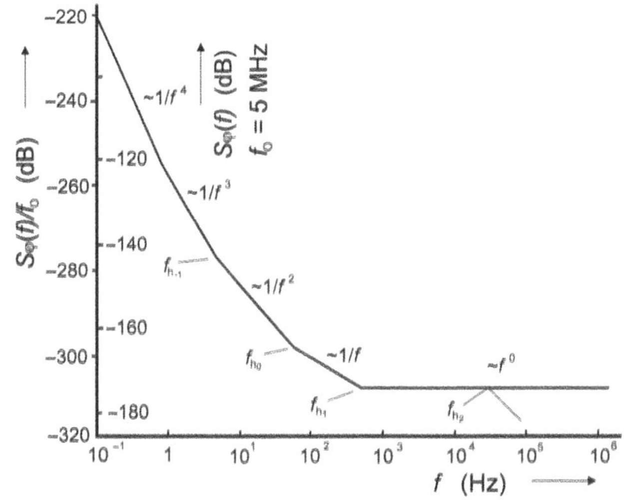

Figure 10.16 Piecewise linearized noise characteristics of an oscillator. *Source:* [6]/*with permission of John Wiley & Sons.*

with phase fluctuations can be described by the relation

$$u(t) = U_0 \sin[\omega_0 t + \phi(t)]. \tag{10.33}$$

Deriving the argument, the instantaneous frequency is obtained

$$\omega_0 + \frac{\mathrm{d}\phi(t)}{\mathrm{d}t} = \omega_0 + \Delta\omega(t). \tag{10.34}$$

Here, $\Delta\omega(t)$ is the parasitic frequency modulation. The relative frequency deviation is defined as

$$y(t) = \frac{\Delta\omega(t)}{\omega_0}. \tag{10.35}$$

Since frequency is the derivative of phase, it applies to the spectral density of relative frequency fluctuations

$$S_y(f) = \left(\frac{f}{f_0}\right)^2 S_\varphi(f). \tag{10.36}$$

A piecewise linear approximation of noise characteristics is then

$$S_y(f) = \sum_{\alpha=-2}^{2} \frac{h_\alpha}{f^{-\alpha}}. \tag{10.37}$$

David Allan [7, 8] proposed the use of variances of pairs of relative frequency deviations to obtain coefficients h_α of piecewise linearized noise characteristics (Eqs. 10.32 and 10.37). Allan's variance of pairs is defined by

$$\sigma^2(\tau) = \frac{1}{2}(\bar{y}_{k+1} + \bar{y}_k)^2, \tag{10.38}$$

where $\bar{y}_k$ is the mean value $y(t)$ per time τ measured in moments t_{k-1}, t_k, t_{k+1}:

$$\bar{y}_k = \frac{1}{\tau}\int_{t_{k-1}}^{t_k} y(t)\mathrm{d}t. \tag{10.39}$$

Conversion relations between variances of pairs $\sigma^2(\tau)$ and coefficients h_α are given in Table 10.1.

Conversion relations are valid the opposite way as well. This means that if spectral power density of the phase noise is measured at suitable points (a piecewise linearized noise characteristic of the oscillator can be computed), Allan's variance for the given interval τ can be determined. We considered this fact when determining Allan's variance of our GPS-GPG24 standard (we only know measured values of the phase noise; see Fig. 10.17). The corresponding Allan's variances are shown then in Figure 10.18. Table 10.2 was completed in this way with the needed information for the design of the satellite receiving station (see Fig. 10.19). For all synthesized frequencies, the GPG24 standard is used. As is obvious from Table 10.2, the GPS standard shows very good long-term stability and a higher dispersion (short-term stability) than the rubidium oscillator (see Fig. 10.20). Therefore, an Rb oscillator disciplined by the GPS standard is considered. Table 10.2 also shows that a good control oscillator can be a crystal-controlled oscillator (XO), especially if placed in a thermostat.

10.3.8 Circuitry of Modern PLL Synthesizers

The circuitry of PLL synthesizers has evolved intensively, and has always been technology dependent. A detailed listing is beyond the scope of this book. However, the circuitry of today's synthesizers is discussed in a few notes.

Table 10.1 Conversion relations between h_α and $\sigma^2(\tau)$. *Source: 6. Kroupa, V.F. (1995). Theory of Phase-Locked Loops and their Application in Electronics (in Czech): 1995. Prague, CR: Academia. ISBN 80-200-0531-5.*

$S_y(f)$	$\sigma^2(\tau)$	Slope (dB/dec)
$h_2 f^2$	$\frac{3h_2 f_H}{(2\pi\tau)^2}$ $(2\pi\tau f_H >> 1)$	−20
$h_1 f$	$\frac{h_1}{(2\pi\tau)^2}[1.038 + 3\ln(2\pi\tau f_H)]$	−20
h_0	$\frac{h_0}{2\tau}$	−10
$\frac{h_{-1}}{f}$	$2h_{-1}\ln 2$	0
$\frac{h_{-2}}{f^2}$	$\frac{(2\pi)^2\tau h_{-2}}{6}$	+10

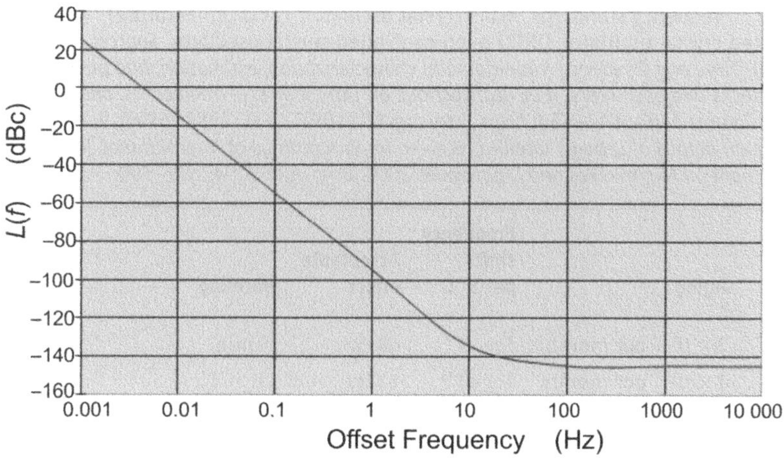

Figure 10.17 Measured noise characteristics of the GPS GPG24 standard, $L(f) \equiv N_\nu(f)$.

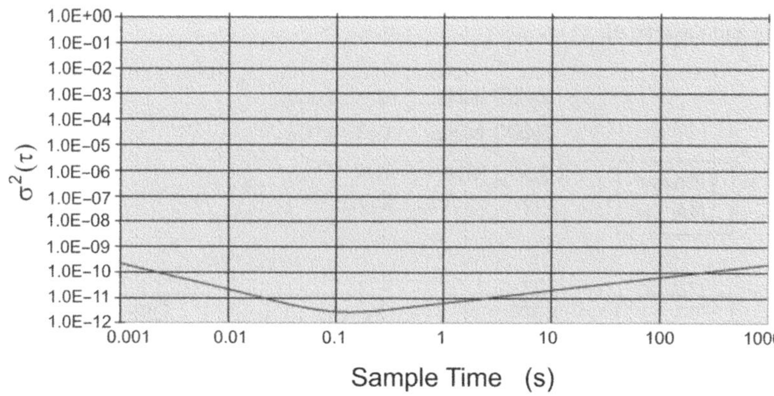

Figure 10.18 Allan's variances of the GPG24 standard computed from the approximated noise characteristics.

10.3.9 Oscillators

Each PLL synthesizer contains at least two oscillators—a master oscillator and a controlled one. Both oscillators significantly influence the quality of generated signals. Requirements imposed on oscillators are different. Low phase noise is very important for a controlled oscillator (VCO) as well, but the requirements are not as strict as for a master oscillator. VCO requires the conversion conductance K_0 corresponding to the range of control voltage u_v and the desired range of capture. VCO operates at high to microwave frequencies. Here, the noise characteristics of oscillators and high-frequency stability are difficult to be achieved. If the synthesizer

Table 10.2 Frequency standards. XO is crystal oscillator, TXCO temperature-compensated crystal oscillator, OXCO oven-controlled crystal oscillator. *Source: 8. ALLAN, D.W. (1987). Time and frequency (time-domain) characterization, estimation, and prediction of precision clocks and oscillators. IEEE Transactions on Ultrasonics, Ferroelectrics, and Frequency Control 34(6): 647–654. https://doi.org/10.1109/T-UFFC.1987.26997; 9. KASAL, M. (2004). Design details of L-band satellite receiver. In: Proceedings of International Travelling Summer School on Microwaves and Lightwaves 2004, Brno, July 2004, 289–300.*

Type	Aging	Frequency shift per 5 °C	Acceptable offset	Warming	Allan's variance (τ = 1 s)
XO	3×10^{-7} per month	5×10^{-6}	60 Hz	30 min	1×10^{-9}
TCXO	1×10^{-7} per month	1×10^{-6}	13 Hz	3 h	1×10^{-10}
OCXO	2×10^{-8} per month	5×10^{-10}	150 MHz	3 d	5×10^{-12}
Rubidium	1×10^{-11} per month	5×10^{-12}	400 μHz	4 h	7×10^{-12}
GPS	3×10^{-12} per month	—	—	—	1×10^{-11}
Cesium	Primary standard	3×10^{-12}	6 μHz	45 min (21.5 ns h^{-1})	1×10^{-11}

Source: Allan and Kasal [8, 9].

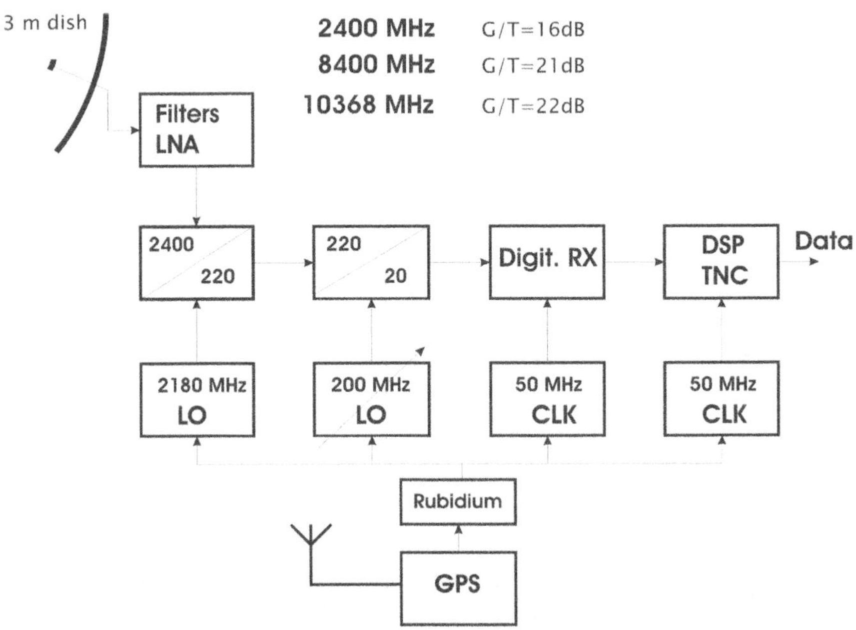

Figure 10.19 Experimental receiving laboratory ESL at the Department of Radio Electronics, Faculty of Electrical Engineering and Communication, Brno University of Technology.

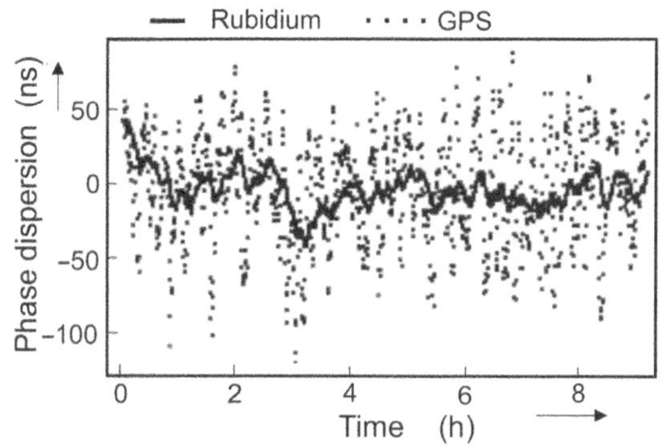

Figure 10.20 Difference of medium-term stability of GPS and Rb frequency normal.

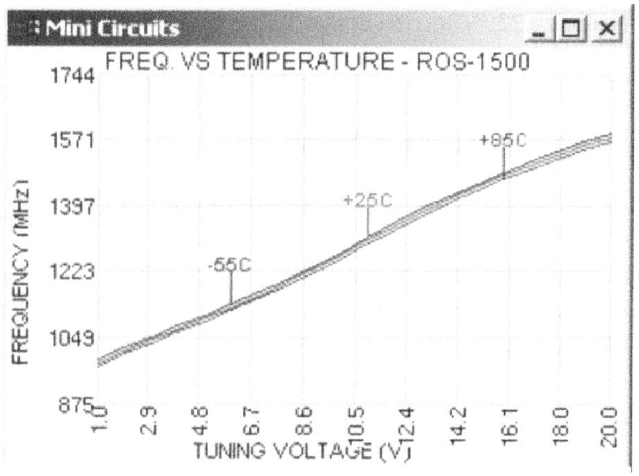

Figure 10.21 Dependence of the frequency of VCO ROS-1500 on the tuning voltage at different temperatures. [10]/ Mini-Circuits.

operates in an environment with a large temperature range (e.g. in a satellite), the temperature stability of the frequency of the free-running oscillator is important as well (see Fig. 10.21).

10.3.10 Phase Frequency Detectors – PFD

In today's PLL synthesizers, standard digital PFDs are almost always supplemented with a charge pump, i.e. a voltage-to-current converter operating in three states with width-modulated pulses (see Fig. 10.22). In a linear model of the PLL, the phase-frequency detector is characterized by the conversion gain K_d(V/rad).

A phase-frequency detector with a charge pump influences noise parameters of the synthesized signal primarily by the size of the reference frequency ΔF. Approximately, $P_{ph} \sim 10\log(\Delta F)$ both outside and inside the loop band.

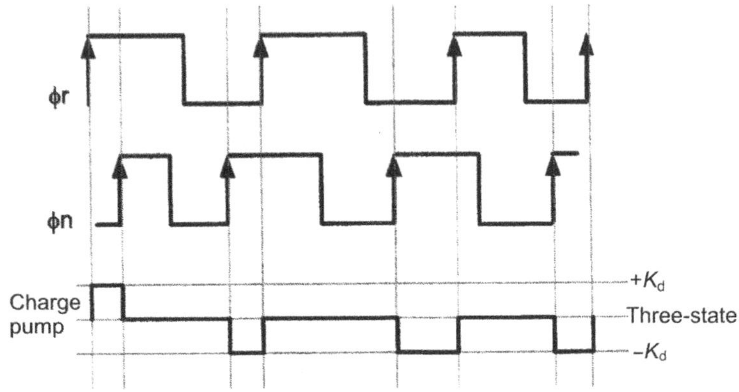

Figure 10.22 Function of the charge pump.

10.3.11 Frequency Dividers

In order to program the output frequency of the synthesizer, a variable frequency divider is included in the feedback branch. Since the output signal of the synthesizer is the input signal of the divider, fixed fast pre-dividers are used at high and very high frequencies. The pre-dividers operate with a double module. The operation of such a divider is shown in Figure 10.23. During a certain part of the cycle, determined by the range of the counter A, the pre-divider divides by the dividing ratio $(P+Q)$. After the counter A is filled in, the quick divider switches to the dividing ratio P. The resulting dividing ratio is given then

$$N = (P + Q)A + P(M - A) = PM + QA \tag{10.40}$$

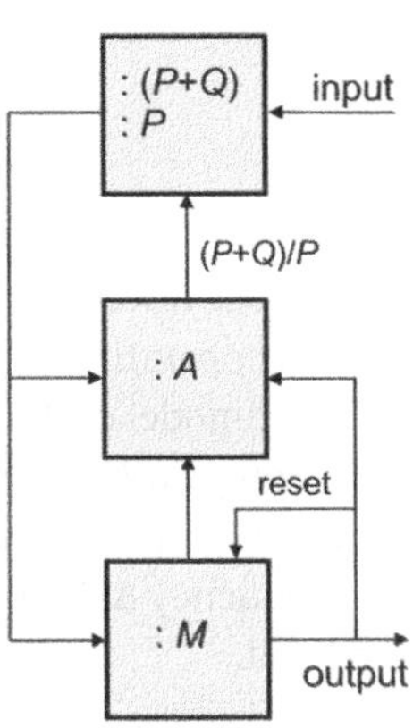

Figure 10.23 Variable divider with two-module fast fixed pre-divider.

Usually, we select $Q = 1$ (pre-divisions 16/17, 64/65, or 20/21) and $M_{min} = A_{max} + 1$. If A changes in the range $0 \le A \le (P - 1)$, the natural numbers N are in the interval

$$PM_{min} \le N \le PM_{max} + A_{max}. \tag{10.41}$$

Inside the PLL band, the contribution of the divider to the phase noise of the PLL is approximately $P_{ph} \sim 20 \log N$.

10.3.12 Loop Filters

Loop filters significantly influence the dynamic behavior of the PLL since they essentially determine the order and type of the PLL. Loop filters are conventional low-pass filters of the first order or the second one (exceptionally a higher order), with current excitation in case of a charge pump. Loop filters are designed as passive or even active ones. Since an active filter in the loop increases noise, there must be a compelling reason for an active solution (e.g. the range of the charge pump is not sufficient to retune the VCO, or due to the phase coherence, the filter is needed to behave as an ideal integrator). Basic relations for a PLL of the second order with a simple RC filter and a ratio RC filter of the first order are shown in Figure 10.24.

(a)

$$\tau_1 = R_1 C_1$$

$$F(s) = \frac{1}{1 + s\tau_1}$$

$$I(s) = \frac{K/\tau_1}{s^2 + s/\tau_1 + K/\tau_1}$$

(b)

$$\tau_1 = C_1(R_1 + R_2), \quad \tau_2 = R_2 C_1$$

$$F(s) = \frac{1 + s\tau_2}{1 + s\tau_1}$$

$$I(s) = \frac{(K/\tau_1)(1 + s\tau_2)}{s^2 + s(1 + K\tau_2)/\tau_1 + K\tau_1}$$

where $K = K_d K_o$

Natural frequency (theory of servomechanisms):

(a) $\omega_n = \sqrt{K/\tau_1}$ (b) $\omega_n = \sqrt{K/\tau_1}$

Attenuation:

(a) $$\zeta = \frac{\omega_n}{2}\frac{1}{K} = \frac{1}{2}\frac{1}{\sqrt{K\tau_1}} = \frac{1}{2}\frac{1}{\omega_n \tau_1}$$

(b) $$\zeta = \frac{\omega_n}{2}\left(\tau_2 + \frac{1}{K}\right)$$

Figure 10.24 Two simplest types of second-order phase-lock filters: (a) Simple RC filter; (b) Proportional RC filter.

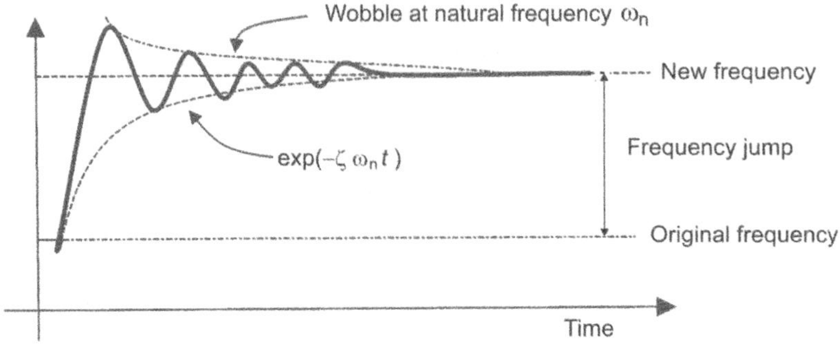

Figure 10.25 Transient phenomenon when switching the phase-locked loop from one frequency to another one.

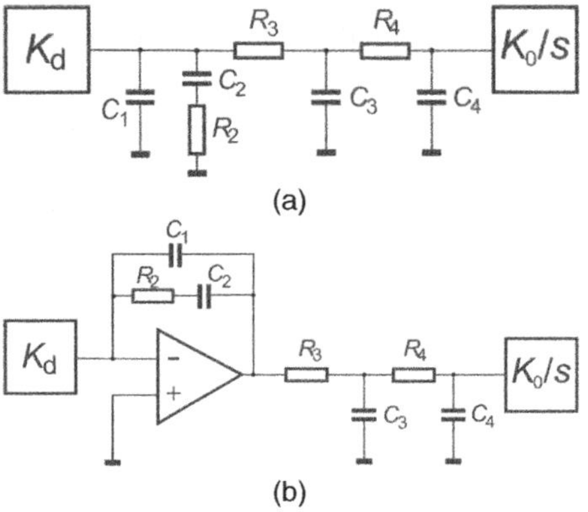

Figure 10.26 Typical filter configurations of a phase-locked loop: (a) Passive; (b) Active.

Definitions of natural frequency ω_n and attenuation ζ are very important since they determine transient phenomena when switching from one frequency to another one (see Fig. 10.25). The choice of time constants strongly determines the bandwidth of the loop ω_p, and together with $|F(j\omega)|$ depends on all other parameters K_d, K_o, ΔF and N for a given application. There are many analyses focused on a more or less optimal calculation of time constants and subsequent calculations of values of circuit elements. Here, computer simulations are widely used allowing to include all essential influences in the calculation of the linearized model (see Fig. 10.26). The presented simulations correspond to configurations of the loop filter according to Figure 10.27. Even the original design considers filters of the first and second orders, and the given topology corresponds to the fourth order. In calculations, VCO decoupling circuits have to be included, even if their time constants are relatively small.

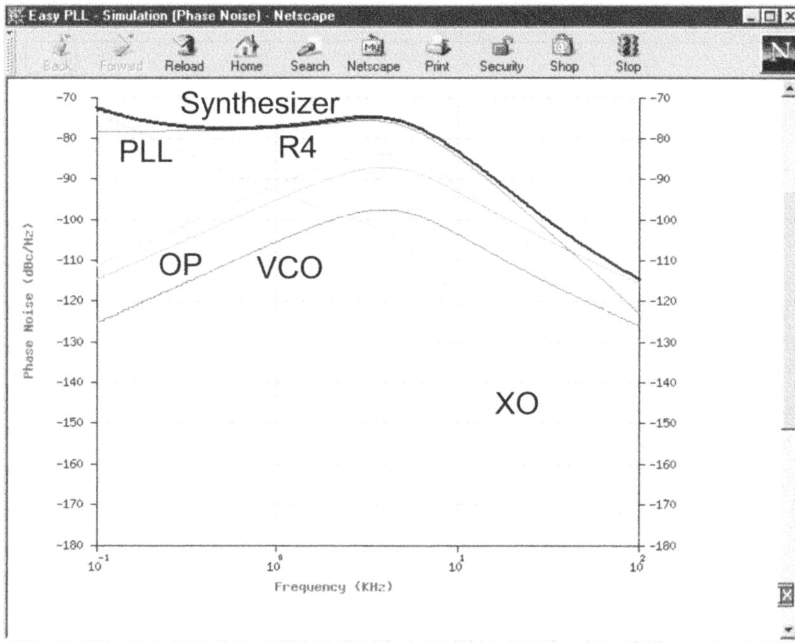

Figure 10.27 Simulation of the phase noise of the synthesizer with marked influence of individual components. *Source:* [11]/*Analog Devices, Inc.*

10.4 Direct Digital Synthesizer with AD9951EP

The synthesizer was developed for an IF transverter from microwave bands so that the accurate (~1 Hz) compensation of Doppler shift in digital transmissions can be allowed using reflection from the lunar surface. The synthesizer allows simultaneous synchronization of a communication device to the atomic frequency normal of 10 MHz (GPS or rubidium). The synthesizer itself consists of an AD9951EP chip with a clock frequency of 400 MHz, which is derived from the frequency normal 10 MHz multiplied by 2. State ports are handled by an ATmega8 processor with appropriate firmware with an access via RS232 standard serial port. The DDS schematic is shown in Figure 10.28. The output spectrum at a generated frequency of 100 MHz is depicted in Figure 10.29. Later, this DDS was used in several specimens to synchronize the voltage-controlled crystal oscillator VCXO (see Fig. 10.30). The DDS in this application operated at 10 times lower frequency.

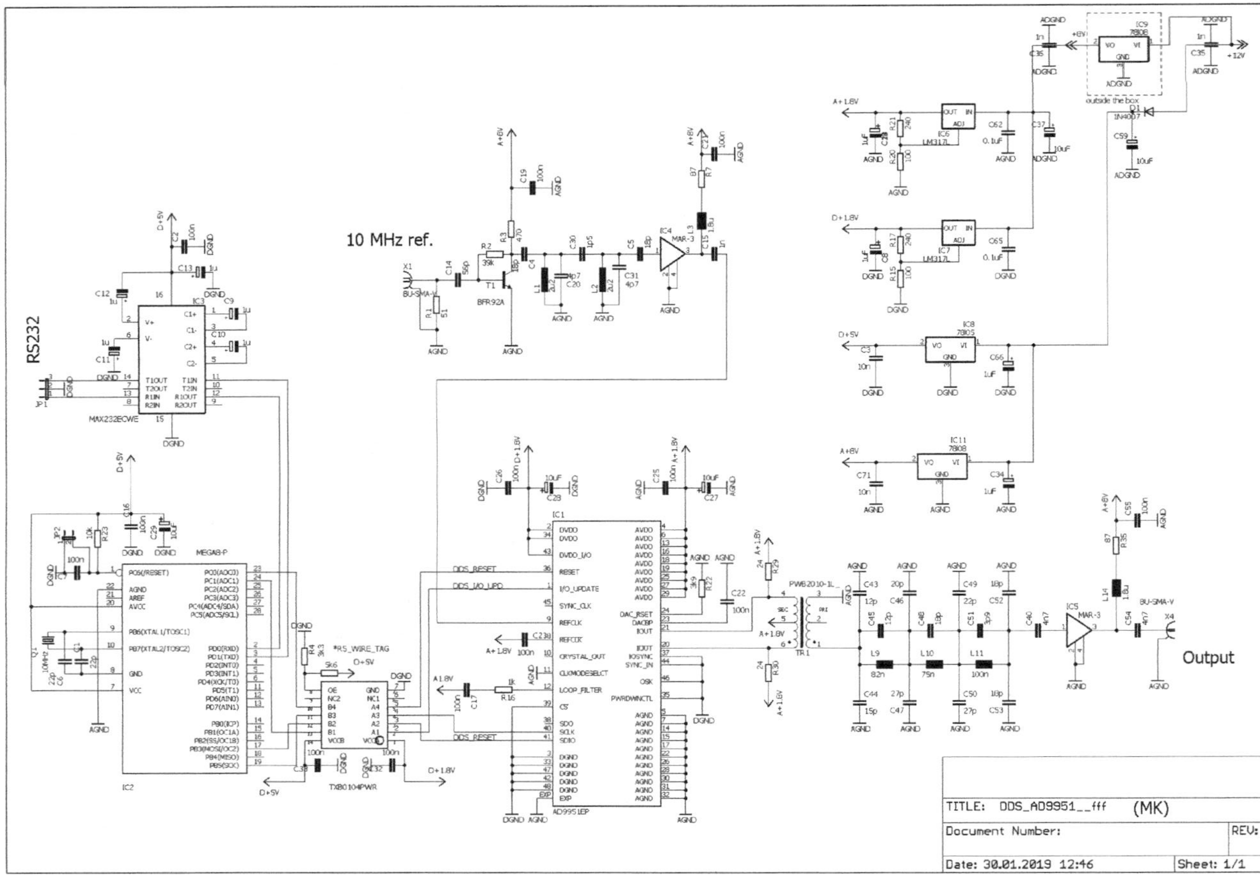

Figure 10.28 Schematics of DDS with AD9951EP.

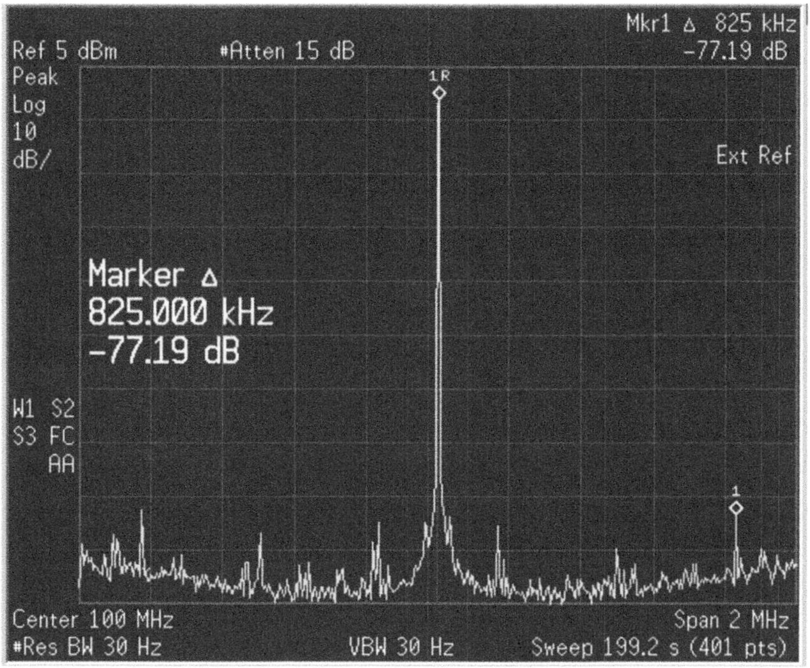

Figure 10.29 The output spectrum of DDS at the generated frequency of 100 MHz.

10.5 Frequency Synthesizer for Microwave Devices

Today's requirements for local oscillators of microwave devices are enormous, especially if weak narrow-band signals are going to be processed. Absolute frequency accuracy has to be comparable to atomic oscillators. The output frequency has to be programmable with a small step. Strong requirements cover phase noise and presence of parasitic components in the output spectrum of the synthesizer. Fortunately, advances in technology of industry leaders (Analog Devices, MAXIM Integrated) allow to design and implement such synthesizers. The chip IO MAX2871 is advantageous since it contains four integrated VCOs that are switched according to scale [12]. This solves the usually critical part of the PLL. Circuits of the synthesizer were supplemented with an electronic switch of an internal and external frequency normal. Figure 10.31 shows that the synthesizer has two outputs. MACOM produces an active doubler with input frequency range of 2.5–6 GHz and output frequency range of 5–12 GHz, delivering up to 16 dBm of output power. The synthesizer was implemented and the multiplier was connected to one output. When controlled by an external normal of 10 MHz

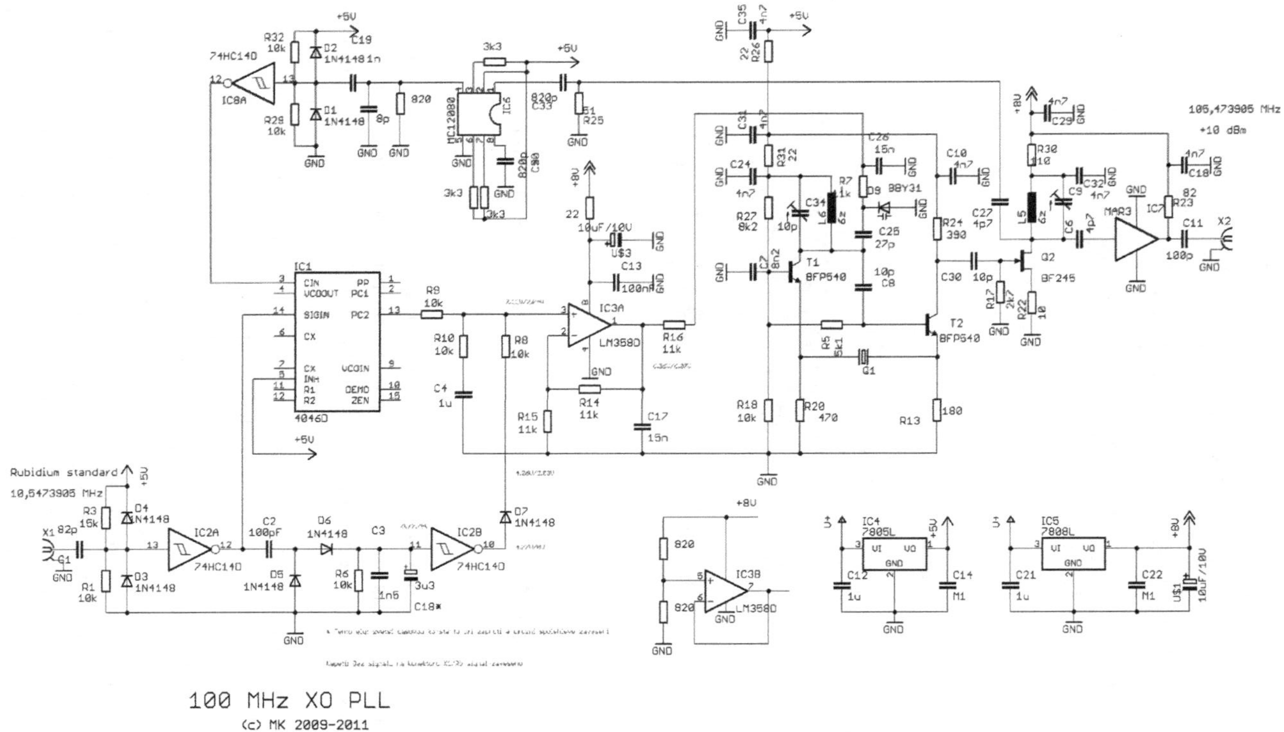

Figure 10.30 VCXO with phase-locked loop synchronized by DDS with AD9951EP. *Source:* [9]/*with permission of Institute of Scientific Instruments of the Czech Academy of Sciences.*

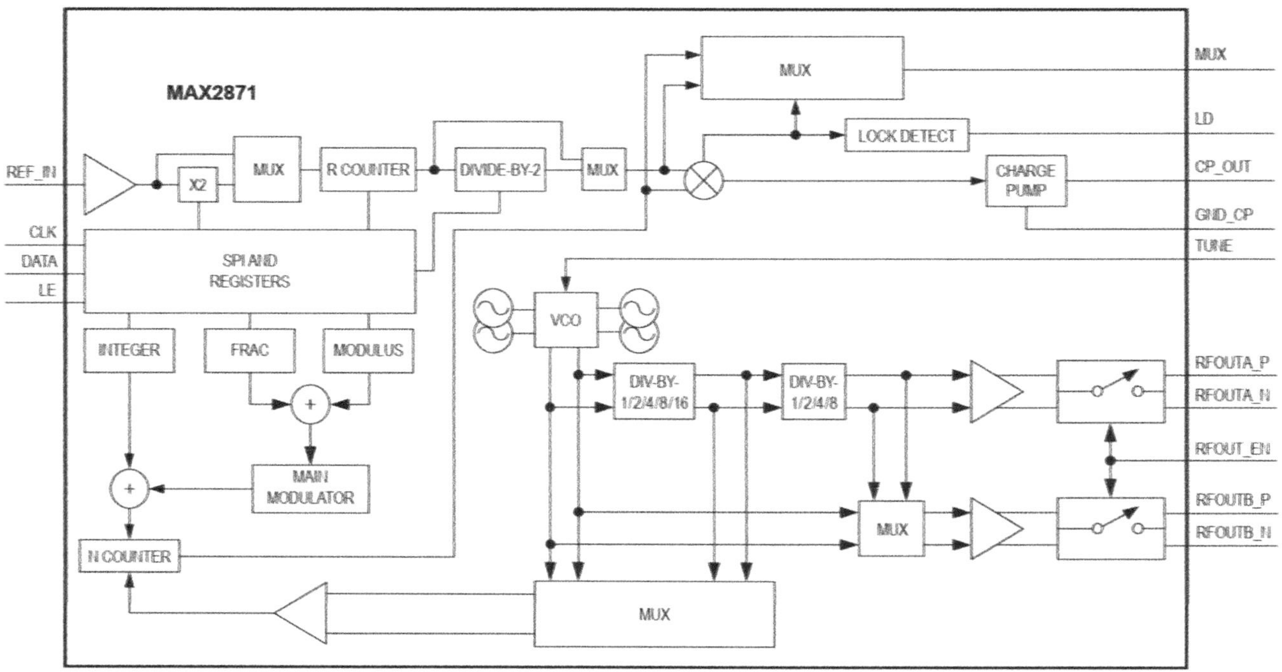

Figure 10.31 Block diagram of a microwave synthesizer up to 6 GHz (the chip MAX2871). *Source:* [12]/*with permission of Analog Devices.*

Figure 10.32 Implementation of a synthesizer MAX2871 with a multiplier XX1002-QH.

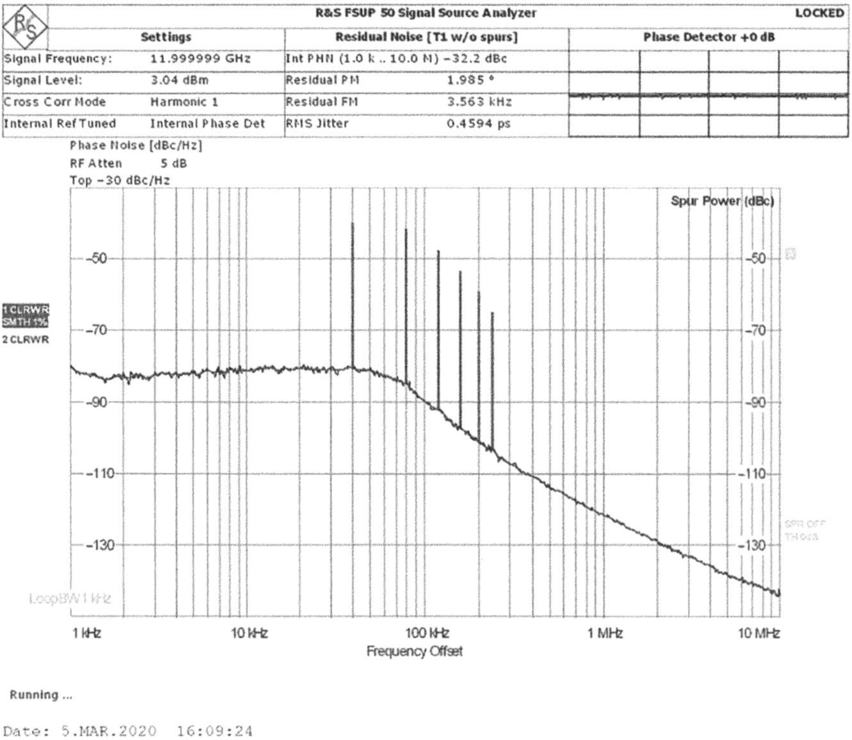

Figure 10.33 Phase noise of a synthesizer MAX2871 with a multiplier XX1002-QH at 12 GHz. *Source:* [13]/*with permission of LYRASIS, Ústřední knihovna VUT*.

(GPSDO), a high-quality synthesizer up to 12 GHz was obtained [13]. The design of the synthesizer is shown in Figure 10.32. Figure 10.33 shows the measured phase noise of the synthesizer at a frequency of 12 GHz.

References

1 Kasal, M. (2005). Frequency synthesis in communication systems of experimental satellite. In: *Scientific Writings of BUT, Habilitation and Inaugural Writings.* 169: 1–35. Brno, CR: VUTIUM. ISBN 80-214-2982-8.
2 Kroupa, V.F. (2003). *Phase Lock Loops and Frequency Synthesis.* Chichester, UK: Wiley. https://doi.org/10.1002/0470014105.
3 Kroupa, V.F. (ed.) (1998). *Direct Digital Frequency Synthesizers.* New York: Wiley. ISBN 978-0-780-33438-0.
4 Rohde, U.L. (1982). *Digital PLL Frequency Synthesizers –Theory and Design.* London: Prentice-Hall International. ISBN 9780132142397.
5 Banerjee, D. (2017). *PLL Performance, Simulation and Design.* 5th ed. Dog Ear Publishing. https://www.ti.com/lit/ml/snaa106c/snaa106c.pdf (accessed 7 August 2024).
6 Kroupa, V.F. (1995). *Theory of Phase-Locked Loops and their Application in Electronics* (in Czech): 1995. Prague, CR: Academia. ISBN 80-200-0531-5.
7 Allan, D.W. (1966). Statistics of atomic frequency standard. *Proceedings of the IEEE* 54(2): 221–231. https://doi.org/10.1109/PROC.1966.4634.
8 Allan, D.W. (1987). Time and frequency (time-domain) characterization, estimation, and prediction of precision clocks and oscillators. *IEEE Transactions on Ultrasonics, Ferroelectrics, and Frequency Control* 34(6): 647–654. https://doi.org/10.1109/T-UFFC.1987.26997.
9 Kasal, M. (2004). Design details of L-band satellite receiver. In: *Proceedings of International Travelling Summer School on Microwaves and Lightwaves 2004,* Brno, July 2004, 289–300.
10 Mini-circuits. (2020). *Local Oscillators.* https://www.minicircuits.com/WebStore/Oscillators.html (accessed 7 August 2024).
11 Analog Devices. (2024). *ADIsimPLL Design Tool.* https://www.analog.com/en/lp/resources/adisimpll.html (accessed 7 August 2024).
12 Analog Devices. (2021). *MAX 2871 Datasheet.* https://www.analog.com/en/products/max2871 (accessed 7 August 2024).
13 Klapil, F. (2020). Frequency synthesizer for microwave communication systems. Diploma thesis. Brno University of Technology. http://hdl.handle.net/11012/189303 (accessed 7 August 2024).

11

Microwave Transmitters

Suitable microwave transmitters are needed to generate signals that are later received and processed as weak signals. This is the case when communicating by reflection from the lunar surface (Earth-Moon-Earth, EME). The signal delay of about 2.5 seconds allows even to measure its own response. When communicating with satellites, a transmitter is needed to transmit signals or to command satellites. The transmitter, therefore, belongs to the standard equipment of an experimental workplace.

For powers in the order of tens of watts in microwave bands, vacuum tubes were mainly used recently (TWT,[1] or reflex klystrons). However, the development of semiconductor technologies has fundamentally changed the situation. The range of hundreds of MHz to lower units of GHz is completely dominated by LD MOSFET transistors, which are very stable from the viewpoint of setting the operating point, have a high gain comparable to vacuum tubes, and show a good efficiency. And types for higher voltage (50 V) especially show a good linearity as well. For frequencies of tens of GHz, power GaAsFET transistors (see Fig. 11.1) have been developed. Since their gain is relatively low, those transistors are mostly operated in class A, resulting in high power consumption and low efficiency. The most advanced elements for these bands are based on gallium nitride (GaN). Those elements show a higher amplification and can operate in the AB class with a higher efficiency. Those elements have an excellent thermal stability, and working with them is similar to above-mentioned LD MOSFETs [2].

In the following text, we present solutions of some problems we encountered during the design and implementation of transmitters and their power modules.

1 Traveling wave tube (TWT).

Reception of Weak Radio Signals from Space, First Edition. Miroslav Kasal.

Figure 11.1 GaAsFET FLM0910-12F for 9–10 GHz band and 12 W power. *Source: Kasal* [1].

11.1 UHF Band Power Amplifier 300 W

For the UHF band, a 300 W power amplifier, LD MOSFET MRF6V4300NBR1, was used at the operating point $V_{DD} = 50\,V$, $I_{DQ} = 0.9$ A for class AB at load power $P_{out} = 300\,W$ [3]. The manufacturer provides the following impedances for power matching at the input and the output (see Table 11.1).

The matching was modeled by Ansoft Designer. Figure 11.2 shows the model and the analysis of the output circuitry in the range 350–500 MHz. The structure of the amplifier (including the microstrip lines) is shown in Figure 11.3. A 30 mils thick ARLON CuClad250GX with relative permittivity $\varepsilon_r = 2.55$ was used as a substrate.

Cooling the amplifier was identified as a significant problem. Considered applications require continuous full power with duty cycle 1:1 (i.e. 50%). At achieved efficiency 50%, the thermal power to be dissipated is 150 W. The amplifier is therefore multipoint bolted to 10 mm thick copper plate, which is further bolted to a massive aluminum finned cooler with two fans. The use of an aluminum "clamp" dissipating heat from the top of the transistor case was absolutely essential (see Fig. 11.3).

Table 11.1 Impedance for power matching of the MRF6V4300NBR1 transistor at the input and the output.

	$V_{DD} = 50\,V_{DC}$, $I_{DQ} = 900$ mA, $P_{out} = 300\,W_{CW}$	
f	Z_{source}	Z_{load}
MHz	Ω	Ω
450	0.39 + j 1.26	1.27 + j 0.96

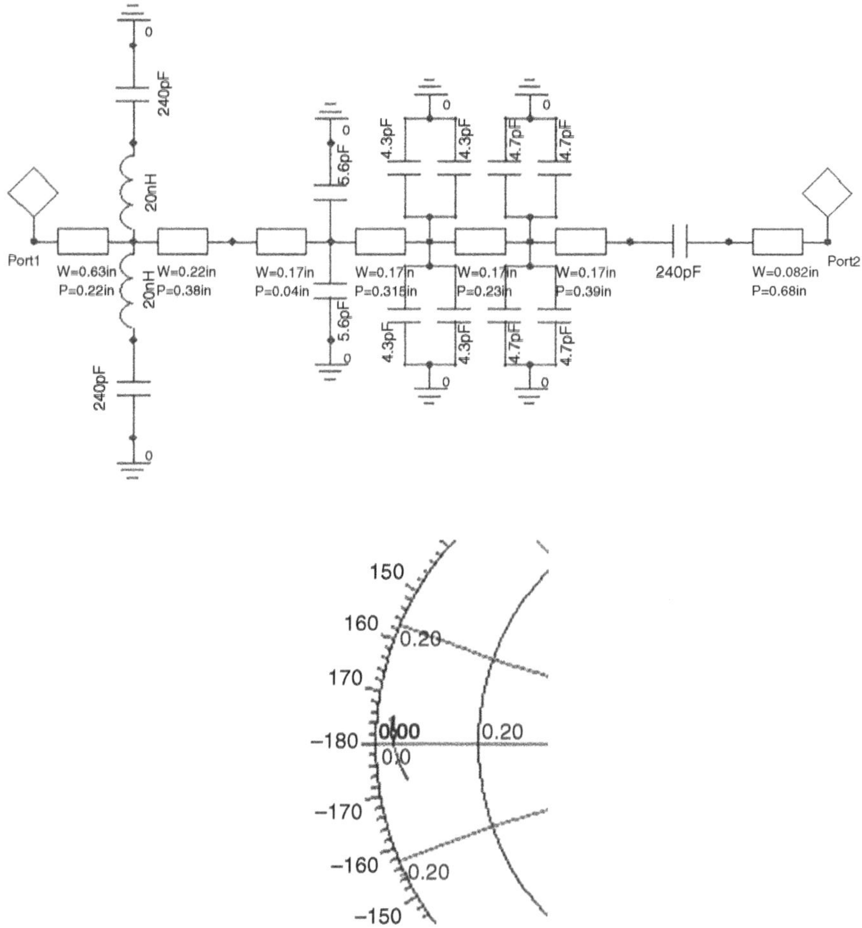

Figure 11.2 Simulation of matching at the LD MOSFET output and its calculation ($Z_0 = 50\,\Omega$).

At the output of such an amplifier, an effective low-pass filter has to be used to suppress the second and third harmonics mainly (see Fig. 11.4a). Capacitors are implemented in form of "facets" on the substrate. The low-pass filter has to transmit the generated power with minimal losses (see Fig. 11.4b). The design of the 300 W amplifier for the UHF band is shown in the photo in Fig. 11.5.

Figure 11.3 Structure of a 300 W UHF power amplifier.

(a)

Figure 11.4 (a) Low-pass filter at the transmitter output; (b) Its insertion loss at 435 MHz does not exceed 0.1 dB.

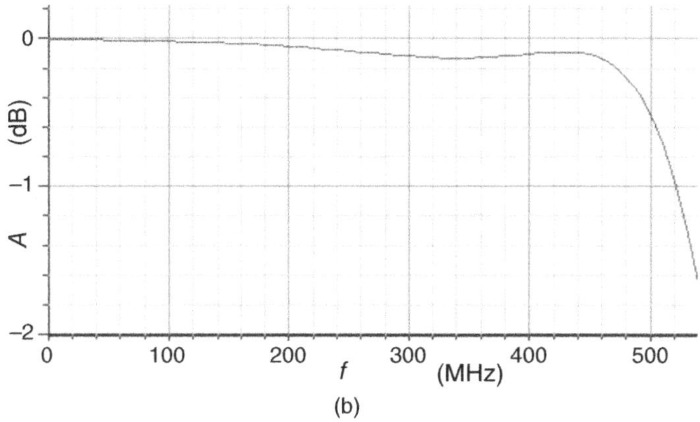

Figure 11.4 (*Continued*)

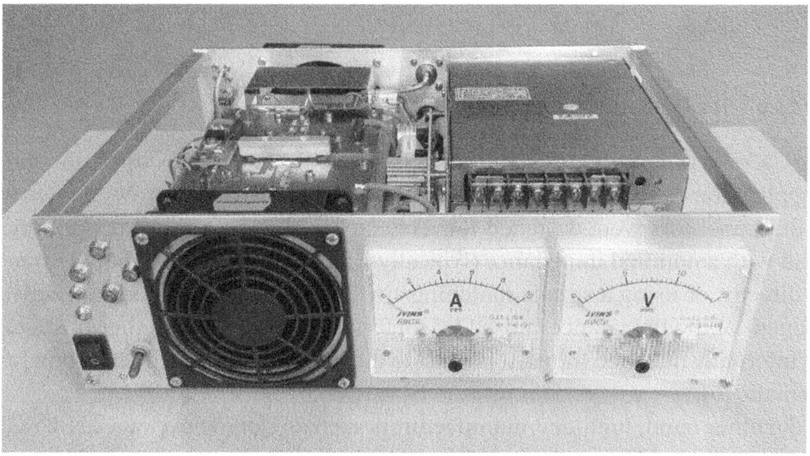

Figure 11.5 The complete design of a 300 W UHF amplifier, including the power supply and cooling system. *Source (photo): radio.feec.vutbr.cz.*

11.2 X-Band Power Amplifiers

At microwave frequencies, the design of power amplifiers (as generally nonlinear circuits) is difficult. Most power amplifiers operate in a linearized mode from the viewpoint of modulation envelope transmission. Therefore, ways of effective designs are still under investigation. A sufficiently precise description of the active element of the amplifier is the basis. Hence, methods of measuring the so-called large-scale s-parameters [4] have been developed. Those methods, however, correspond to a narrow frequency band and a single excitation state of the amplifier. Models based on X-parameters are more universal. The models, including the equipment for their measurement, were designed by Keysight (Agilent) [5]. Today,

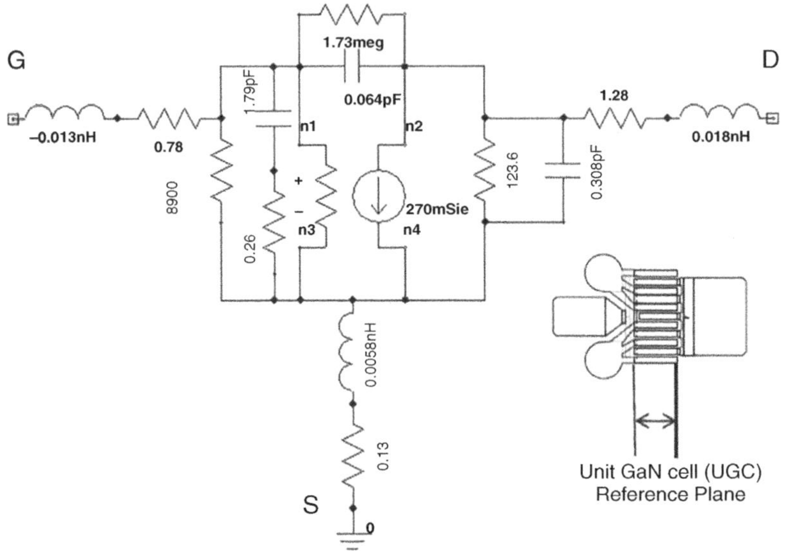

Figure 11.6 The linearized model of the TGF2023-2-01 transistor.

power GaAsFETs (and often LD MOSFETs as well) are manufactured for specific applications and are internally matched in the relatively narrow frequency band the transistors were designed for. This means that both the generator and the load with a nominal impedance (typically 50 Ω) can be connected to the transistor directly, or using a simple matching circuit, which compensates a reactive component of the impedance caused by soldering the transistor to the substrate. Using internally matched transistors outside the recommended frequency band is problematic [6].

On the other hand, high-end manufacturers such as Qorvo develop GaN-based wideband chips that have a reasonable gain in the whole band, e.g. TGF2023-2-01 is 1.25 mm GaN on SiC HEMT providing power up to 38 dBm (6 W) from DC to 18 GHz [7]. The manufacturer provides an equivalent circuit of a linearized model (see Fig. 11.6), which can be used for calculating the s-parameters (see Fig. 11.7).

With simple external circuits, matching can be designed at 10.4 GHz both at the generator and the load (see Fig. 11.8). Calculated s-parameters (see Fig. 11.9) show that resulting characteristics of the amplifier correspond to expected parameters in terms of small-signal gain and matching on both ports [8]. The immittance stability criterion can be applied as well when $20 \log |s_{11}| < 0$ and $20 \log |s_{22}| < 0$ over the entire frequency range. With large signals, stability cannot be analyzed that way. However, the instability of the amplifier usually starts with small signals, so this criterion can be applied. Figure 11.10 shows the equivalent circuit of TGF2023-2-01 including the matching at the input and output.

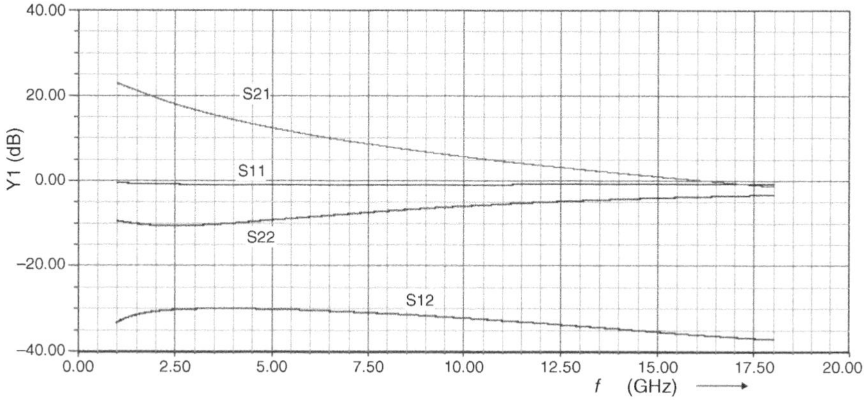

Figure 11.7 Calculated s-parameters of the TGF2023-2-01 transistor.

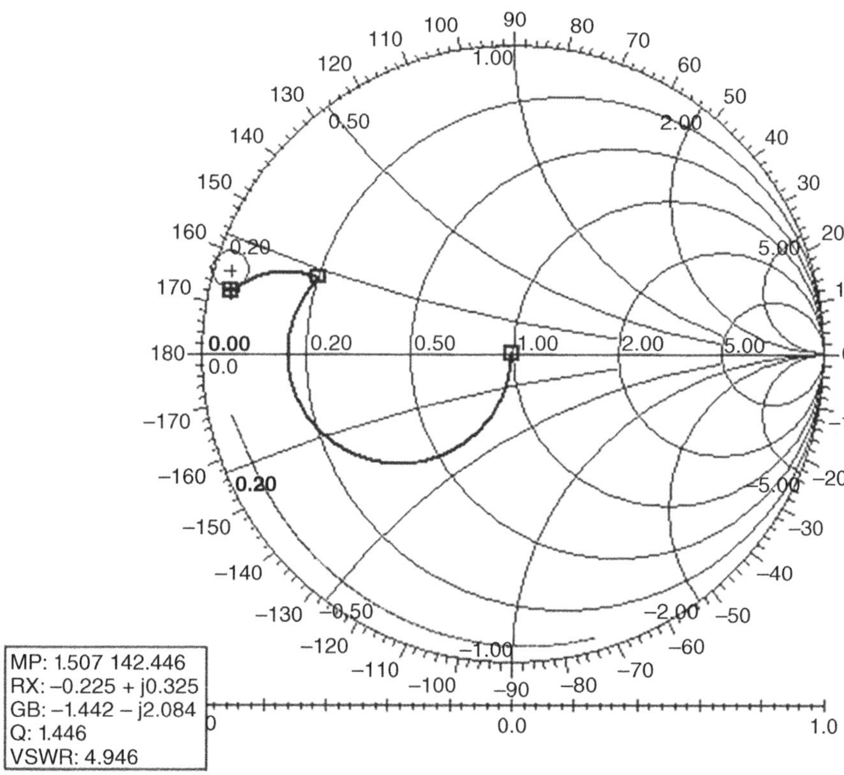

Figure 11.8 Matching at the input of the TGF2023-2-01 transistor. The arc from the center corresponds to a longitudinal section of the transmission line that is 120 ° long and is followed by a grounded stub.

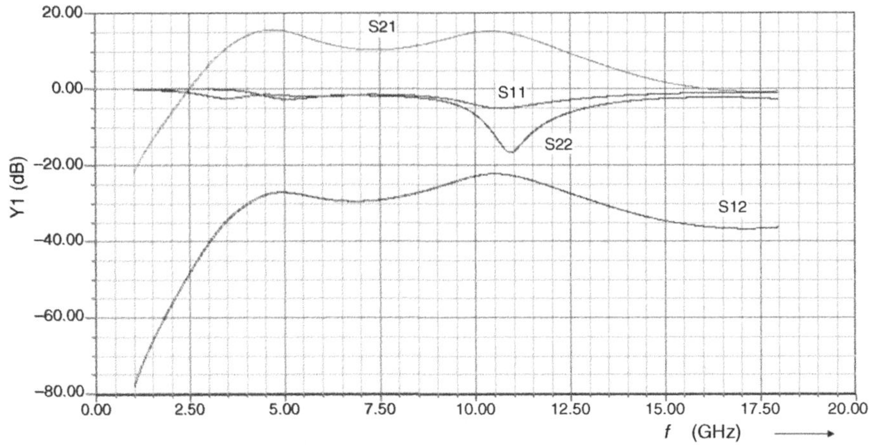

Figure 11.9 Calculated small-signal s-parameters of the amplifier after indicated matching of the TGF2023-2-01 input and output at 10.4 GHz.

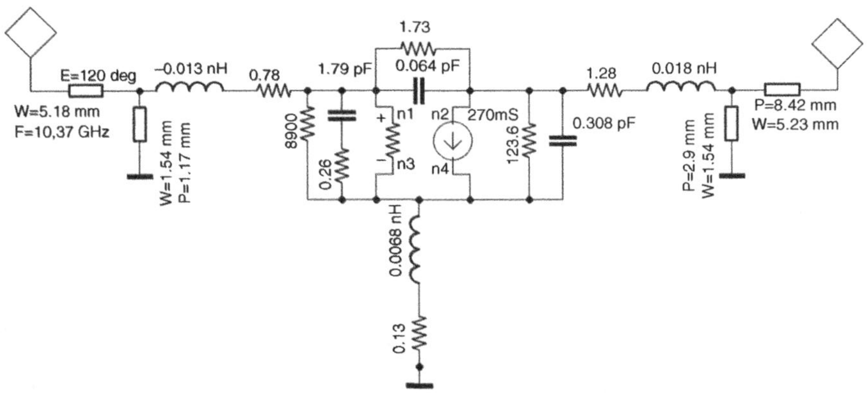

Figure 11.10 Equivalent circuit of TGF2023-2-01 including the matching at the input and output.

11.2.1 Single-Stage X-Band 4 W Power Amplifier

A similar procedure was used to design a single-stage amplifier with an internally matched GaAsFET FLM0910-4F. The transistor operates in class A [9]. For the frequency range 9.3–10.7 GHz, the data sheet provides s-parameters at the operating point $V_{DD} = 10$ V, $I_{DQ} = 1.1$ A. After adding power supply circuits, the amplifier model can be assembled (see Fig. 11.11). To achieve optimal matching at 10.37 GHz, stubs were used both at the input and the output (see Fig. 11.11; see also the amplifier photo in Fig. 11.12). Parameters of the model of the matched amplifier are shown in Figure 11.13. The implemented amplifier showed the calculated parameters $P_{out} = 36$ dBm (4 W) at efficiency $\eta = 0.3$ and the intermodulation distortion IMD < 40 dB.

11.2.2 X-Band 20 W Power Amplifier

For experiments with reflections from the Moon, an amplifier with at least 43 dBm (20 W) of power and 20 dB of large-signal gain was needed to be built at a frequency of 10.37 GHz. The concept of a three-stage amplifier with an output stage operating in quadrature was selected [1]. Transistors in the second and third stages are internally matched GaAsFETs. The amplifier is implemented using microstrip technology on the Duroid 5 880 with thickness 20 mils and $\varepsilon_r = 2.2$. The precise design of the 3-dB hybrid coupler at the input and the output of the output stage is a fundamental issue of the given concept (see Figs. 11.14 and 11.15).

In the first stage, a GaAsFET EPA240B-100P from Excelics Semiconductors is used. Next, FLM0910-4F from Eudyna serves as the drive, and two FLM0910-12 are used in the output stage. In order to achieve a good linearity and the desired gain, all the stages operate in class A. The schematics of the amplifier is shown in Figure 11.16 and the implementation in Figure 11.17.

The power $P_{1\,dB} = 43.2$ dBm was achieved at IM3 < 40 dB and gain of 19 dB. The total DC current of the amplifier at 12 V supply is 8.8 A. The total efficiency (of all three stages) is $\eta = 0.2$. In CW operation, a power of 80 W has to be dissipated. For this reason, the amplifier is mounted on a massive cooler with two 80 mm fans. In CW mode, the temperature of the amplifier does not exceed 70 °C. Figure 11.18 shows a photo of the test bed.

11.2.3 X-Band 50 W Power Amplifier

Due to the development of radar technology, multistage high-power monolithic microwave integrated circuits (MMIC) are being developed (mostly GaN-on-SiC based). Those MMICs are enabling the implementation of solid-state power amplifiers (SSPAs) with an output power of tens of watts. Modules of several top companies can be given as examples: Mimix, CREE, and Qorvo (formerly TriQuint) (see Fig. 11.19).

As already discussed, these modules have been developed for radars with adaptive antenna arrays dominantly. Each element of an antenna array is completed by a separate amplifier excited via a phase element. That way, the resulting radiation direction of the antenna can be changed, and a high radiated power can be achieved. Those amplifiers are primarily intended for the pulse mode. This fact has to be taken into account during the design and implementation of the amplifier. Heat dissipation is a major issue of these circuits when operating with a large duty cycle or even CW. To reduce the thermal resistance between

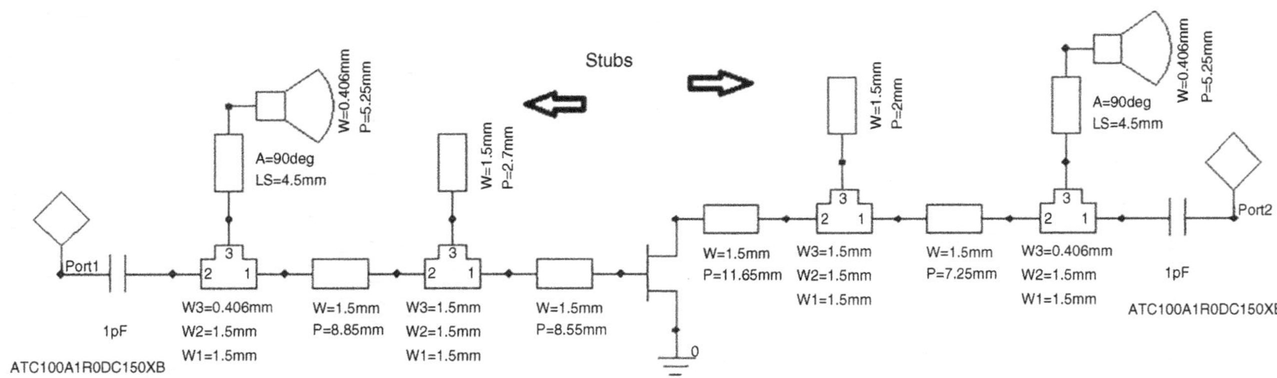

Figure 11.11 The model of the linearized amplifier with FLM0910-4F in class A.

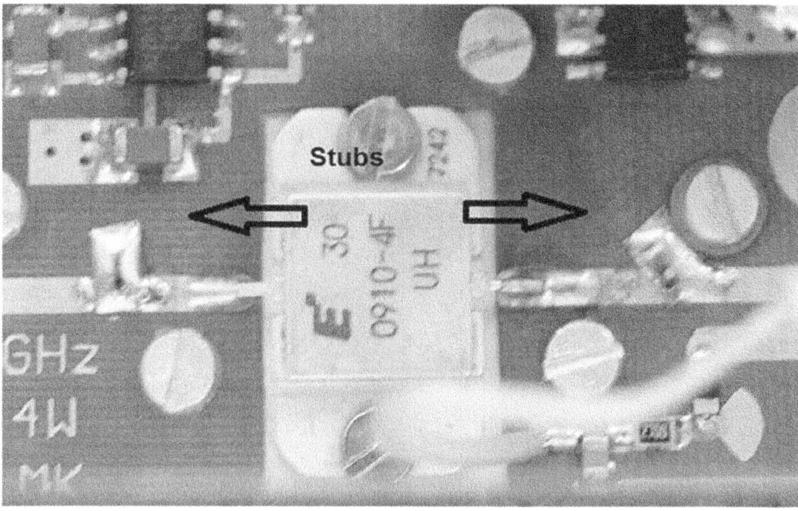

Figure 11.12 Stubs for optimal matching. *Source (photo): radio.feec.vutbr.cz.*

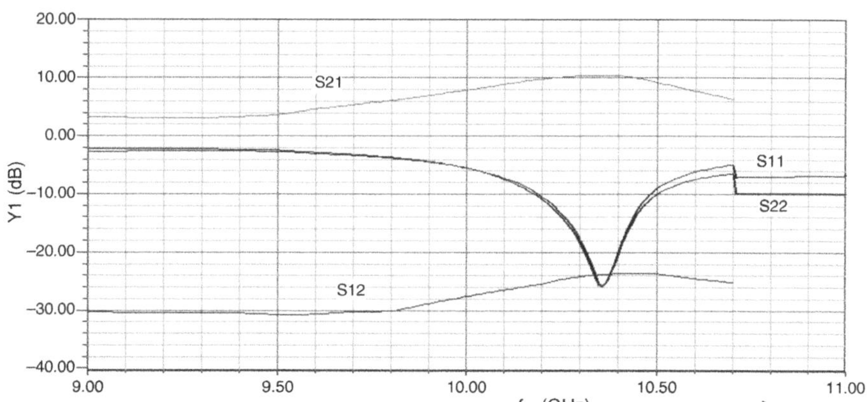

Figure 11.13 Parameters of the amplifier optimized at 10.37 GHz.

the case and the cooler, the case should be soldered to the copper plate with a special indium solder, or a heat-spring made of thin indium foil should be used (see Fig. 11.20). Thanks to the softness, the foil perfectly adheres to both surfaces and reduces the thermal resistance of the transition. Furthermore, a prescribed sequence of connected and disconnected voltages[2] has to be ensured for these

2 In principle, all negative bias voltages have to be connected before positive drain voltages and vice versa when disconnecting.

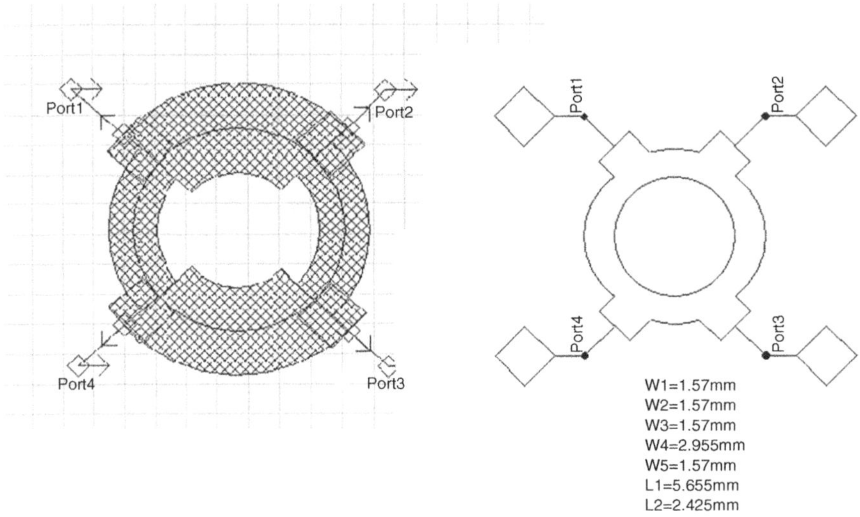

Figure 11.14 Optimized design of 3-dB coupler.

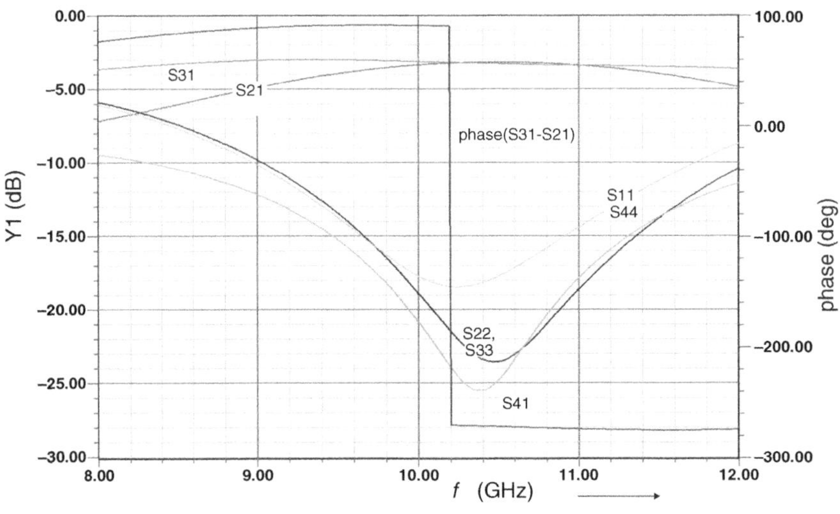

Figure 11.15 Simulated parameters of the optimized 3-dB hybrid coupler.

circuits. Otherwise, there is risk of damaging a very expensive MMIC. The bed of the TGA2312-FL module is made of nickel-plated aluminum. The printed circuit board (PCB) on the RO4003C substrate is soldered to the bed along the entire bottom surface (see Fig. 11.20) to ensure sufficient cooling [6, 13]. Very short input and output microstrips require a small modification for matching at a frequency of 10.37 GHz. Schematics of the amplifier with a driver, a power supply,

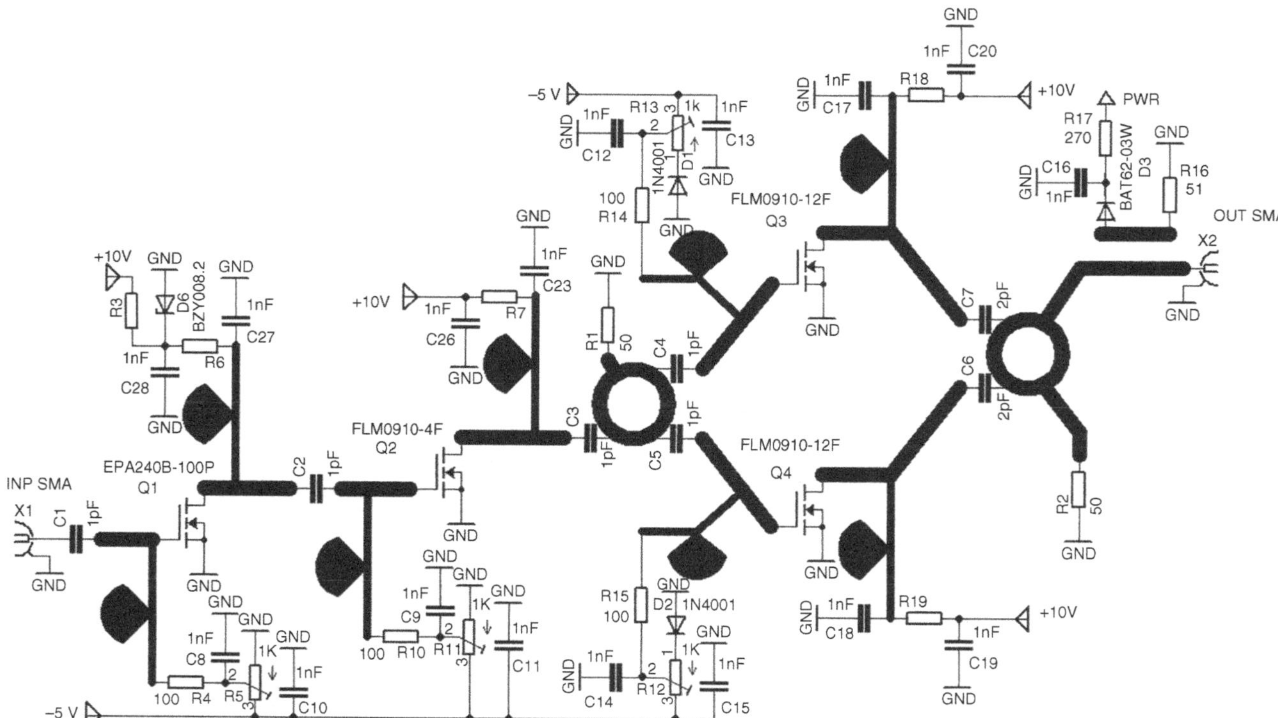

Figure 11.16 Schematics of a 20 W X-band amplifier.

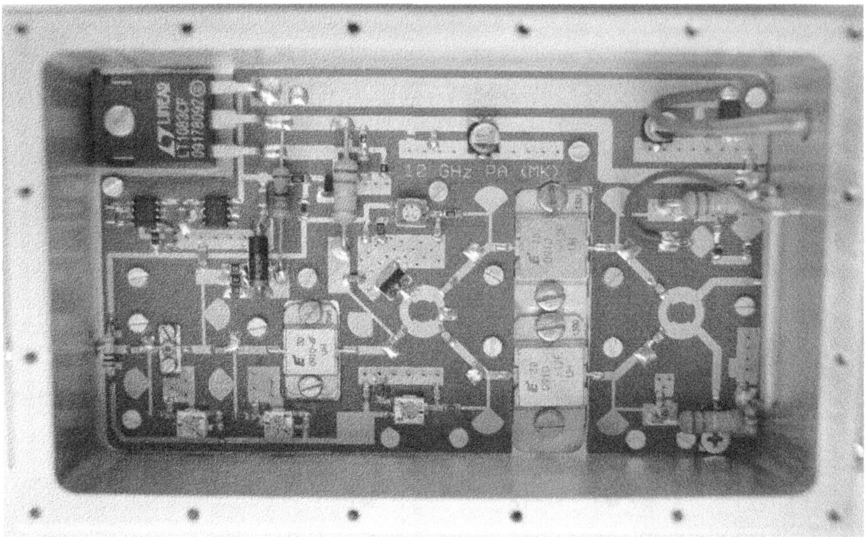

Figure 11.17 Implementation of a 20 W X-band amplifier. *Source (photo): radio.feec.vutbr.cz.*

Figure 11.18 Test bed for microwave power amplifiers. *Source (photo): radio.feec.vutbr.cz.*

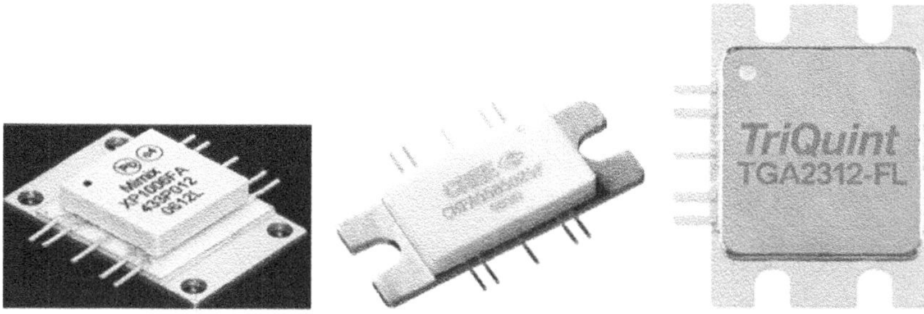

(a) (b) (c)

Figure 11.19 X-band high-power MMICs: (a) Mimix XP1006FA (10 W) [10]/alldatasheet.com; (b) CREE CMPA801B025F (25 W). *Source:* [11]/*Cree LED*; (c) Qorvo TGA2312-FL (60 W). *Source:* [12]/*Qorvo, Inc.*

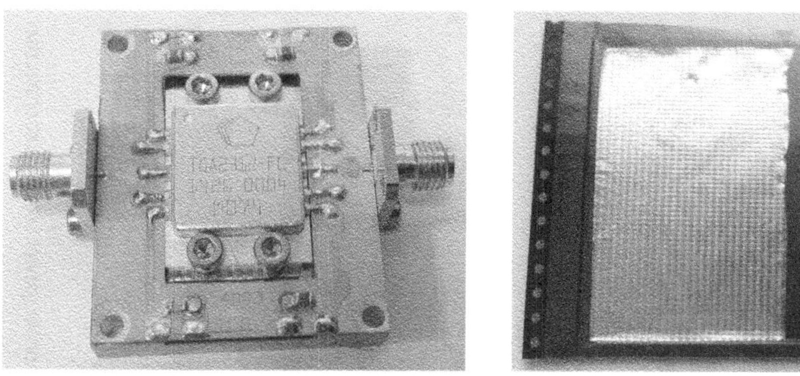

(a) (b)

Figure 11.20 (a) The bed of the TGA2312-FL module with soldered PCB.cz; (b) A sample of indium foil used as a heat-spring. *Source: radio.feec.vutbr.cz.*

and protection circuits is shown in Figure 11.21. The TGA2312-FL contains an internal thermistor, which can be used as an overtemperature protection sensor. With the R17 trimmer, the maximum temperature of the case is set to about 60 °C. The NE555 circuit ensures that the amplifier cannot be turned on for about two minutes when overheated. The driver is similar to the above-described 5 W amplifier. To increase the overall efficiency, the driver is powered by a small 24/12 V converter. The amplifier was built in two pieces, with a more powerful 8 W driver in the second implementation.

11.2.3.1 Calculation of Cooling

In Figure 11.22, the data provided by the manufacturer are given to calculate the thermal balance and amplifier reliability. The power of the amplifier is assumed

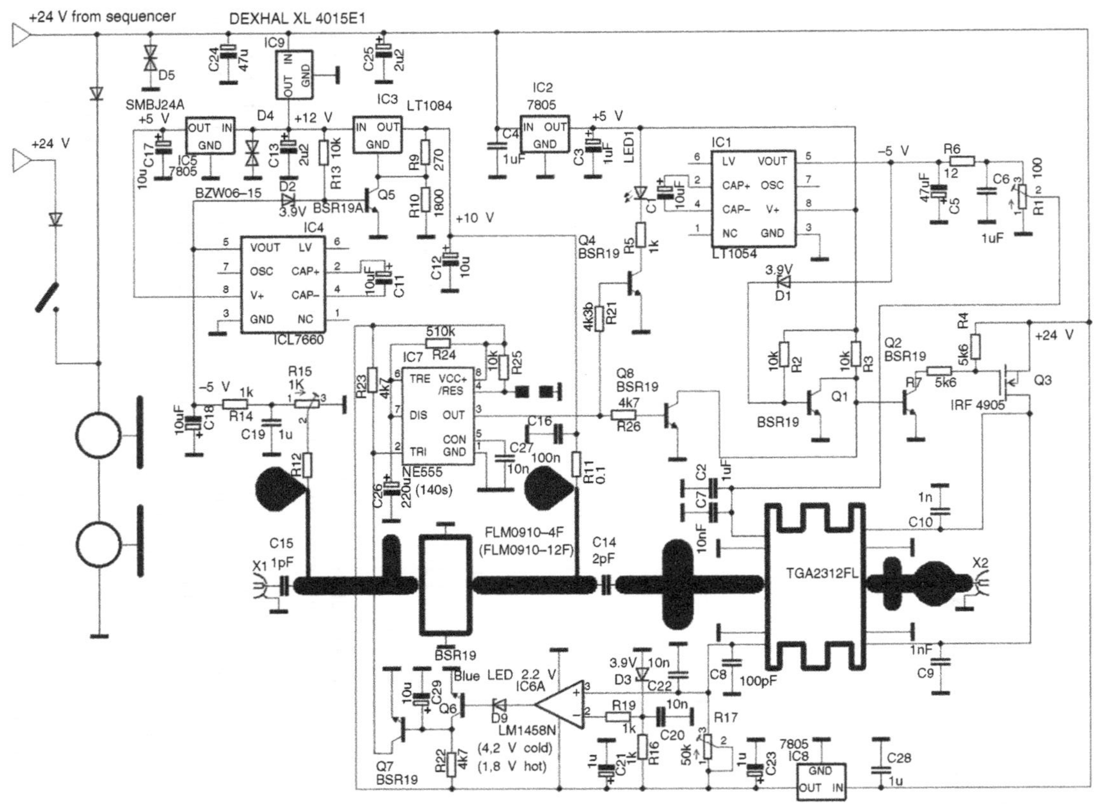

Figure 11.21 Schematics of the 50 W amplifier with a driver, power supply, and protection circuits.

Thermal and Reliability Information

Parameter	Test Conditions	Value	Units
Thermal Resistance, θ_{JC} (Note 1)	Tbaseplate = 85 °C	0.85	°C/W
Channel Temperature, T_{CH} (Without RF Drive)	Tbaseplate = 85 °C, V_D = **24 V**, I_{DQ} = 2400 mA, P_{DISS} = 58 W, Pulsed: PW = 100us, DC = 10%	135	°C
Median Lifetime, T_M (Without RF Drive)		9.75 x 10^10	Hrs
Channel Temperature, T_{CH} (Under RF Drive)	Tbaseplate = 85 °C, V_D = **24 V**, $I_{D\ Drive}$ = 6360 mA, P_{OUT} = 48 dBm, P_{DISS} = 87 W, Pulsed: PW = 100us, DC = 10%	158	°C
Median Lifetime, T_M (Under RF Drive)		7.38 x 10^9	Hrs
Channel Temperature, T_{CH} (Under RF Drive)	Tbaseplate = 85 °C, V_D = **30 V**, $I_{D\ Drive}$ = 6670 mA, P_{OUT} = 48.8 dBm, P_{DISS} = 124 W, Pulsed: PW = 100us, DC = 10%	190	°C
Median Lifetime, T_M (Under RF Drive)		3.12 x 10^8	Hrs

Notes: (1) Thermal resistance measured at back of the package.

Figure 11.22 Data for the thermal balance calculation. *Source:* [12]/*Qorvo, Inc.*

to be $P_{out} = 52$ W. At efficiency $\eta = 0.4$, we get $P_{dc} = 130$ W. The supply voltage is 24 V, and the current is $I_D = 5.4$ A. If the CW mode is considered (the duty cycle 100%), Θ_{JC} is 2x higher (i.e. 1.7 °C/W) than shown in Figure 11.22. If the driver power is 8 W, the heat to be dissipated in the output module is 5.4 A × 24 V + 8 W − 52 W = 86 W. The considered 135 mm long cooler with a profile shown in Figure 11.23 has thermal resistance $\Theta_H = 1.4$ °C/W. After adding two fans, the thermal resistance can be reduced to $\Theta_H^V = 0.24$°C/W. Since the driver with $I_{DB} = 1.6$ A at 24 V is on the same cooler, the temperature rise of the cooler is $\Delta_{TH} = (1.6$ A × 24 V + 86 W) × 0.24 °C/W = 29.8 °C. At a maximum ambient temperature 50 °C,[3] the channel temperature is

$$T_{CH} = 50°\text{C} + 29.8°\text{C} + 86\ \text{W} \times 1.7°\text{C/W} = 226°\text{C}.$$

The value 1×10^6 hours is obtained from the chart in Figure 11.24 for the mean life. This calculation corresponds to the limit mode of the full continuous power of the amplifier and should never be exceeded. Thermal protections are set for significantly lower temperatures.

In case of both implemented amplifiers (see Fig. 11.25), expected parameters were achieved and confirmed by measurements in the laboratory; see Table 11.2. Properties of amplifiers were verified by a series of experiments when communicating with reflections from the Moon. The dependence of the output power on the excitation level of a complete amplifier with an 8 W driver is shown in Figure 11.26. The drain efficiency is given in Figure 11.27.

The first three columms of Table 11.2 shows the input power, namely, the set level of the generator and the level at the input of the amplifier in dBm and in

3 We have to consider that the amplifier is placed in the focal point of a larger parabolic reflector. Even if its reflectivity in the infrared region is small, the amplifier heats up considerably on sunny days.

Table 11.2 Measured parameters of amplifier with 52 W output power.

Gen level	Inp level	Inp level	Meas level	I_d	Input power	Output power	Output power	Drain efficiency	Total efficiency	Gain
dBm	dBm	W	dBm	A	W	dBm	W	%	%	dB
−14	19.10	0.081	6.20	1.90	45.6	36.80	4.79	10	6	17.7
−13	20.10	0.102	7.20	1.90	45.6	37.80	6.03	13	7	17.7
−12	21.15	0.130	8.20	2.00	48.0	38.80	7.59	16	9	17.7
−11	22.15	0.164	9.00	2.20	52.8	39.60	9.12	17	10	17.5
−10	23.20	0.209	10.10	2.40	57.6	40.70	11.75	20	13	17.5
−9	24.20	0.263	10.90	2.60	62.4	41.50	14.13	23	14	17.3
−8	25.20	0.331	11.80	2.80	67.2	42.40	17.38	26	17	17.2
−7	26.20	0.417	12.70	3.10	74.4	43.30	21.38	29	19	17.1
−6	27.20	0.525	13.50	3.30	79.2	44.10	25.70	32	22	16.9
−5	28.30	0.676	14.40	3.60	86.4	45.00	31.62	37	26	16.7
−4	29.40	0.871	15.10	4.00	96.0	45.70	37.15	39	28	16.3
−3	30.50	1.122	15.70	4.30	103.2	46.30	42.66	41	31	15.8
−2	31.60	1.445	16.00	4.50	108.0	46.60	45.71	42	32	15.0
−1	32.70	1.862	16.30	4.80	115.2	46.90	48.98	43	32	14.2
0	33.80	2.399	16.60	4.99	119.8	47.20	52.48	44	34	13.4

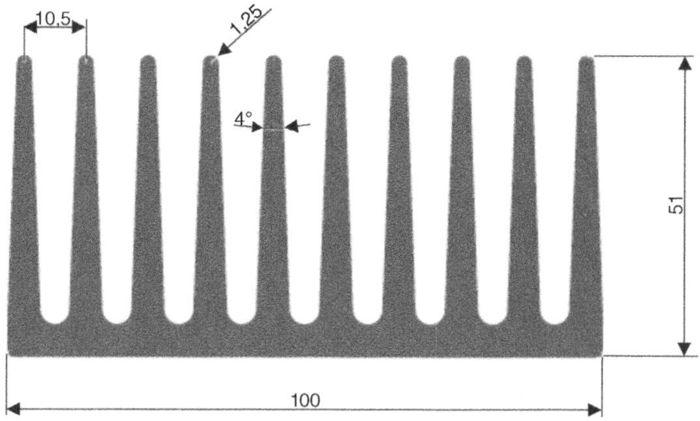

Figure 11.23 Cooler profile.

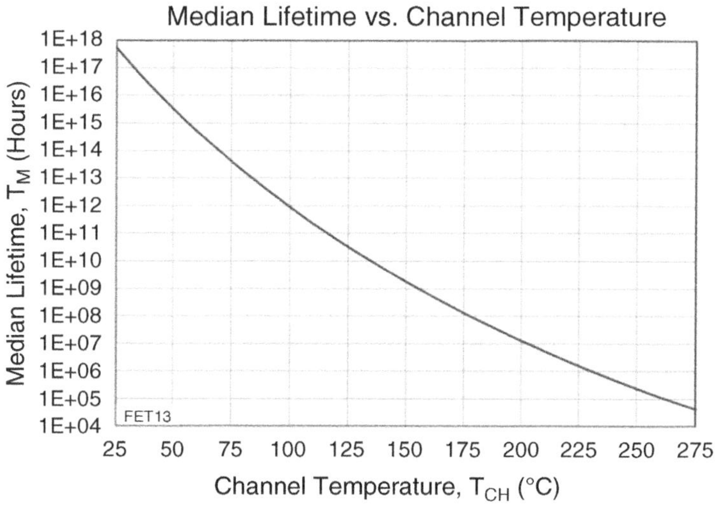

Figure 11.24 Lifetime median of TGA2312-FL versus channel temperature *Source: [12]/reproduced with permission of Qorvo.*

W; the driving amplifier was connected between the generator and the measured amplifier. The following eight columns in Table 11.2 show the measured and calculated values: (1) the level measured on a load with attenuation of 30.6 dB; (2) the direct current of collector (drain); (3) the input power; (4) the output power (dBm); (5) the output power (W); (6) the efficiency of collector (drain); (7) the overall efficiency; (8) the amplifier gain.

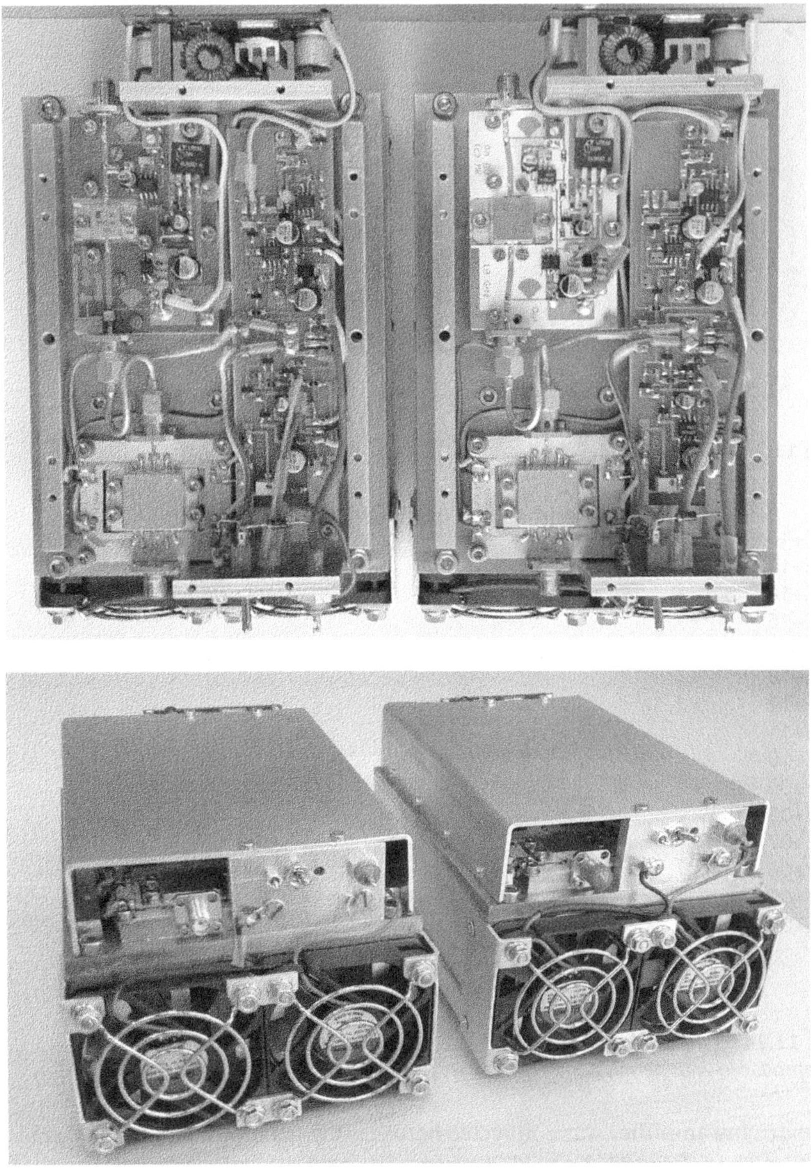

Figure 11.25 The design of amplifiers with TGA2312-FL with powers of 42 and 52 W in the 10 GHz band. *Source: radio.feec.vutbr.cz.*

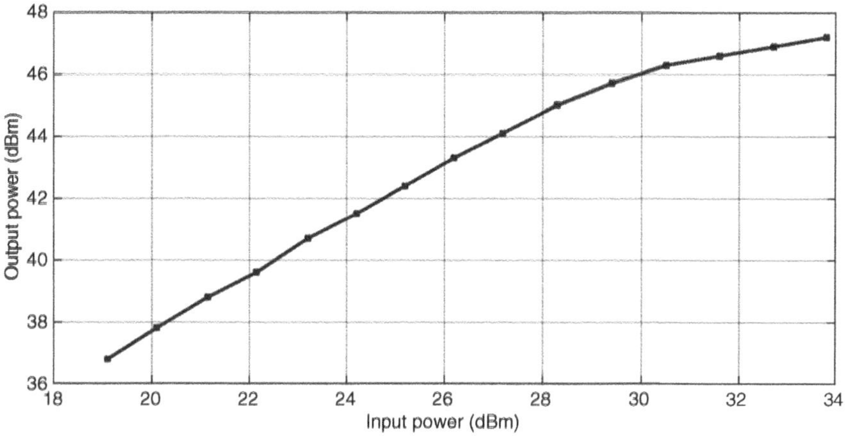

Figure 11.26 Dependence of the output power on the excitation level.

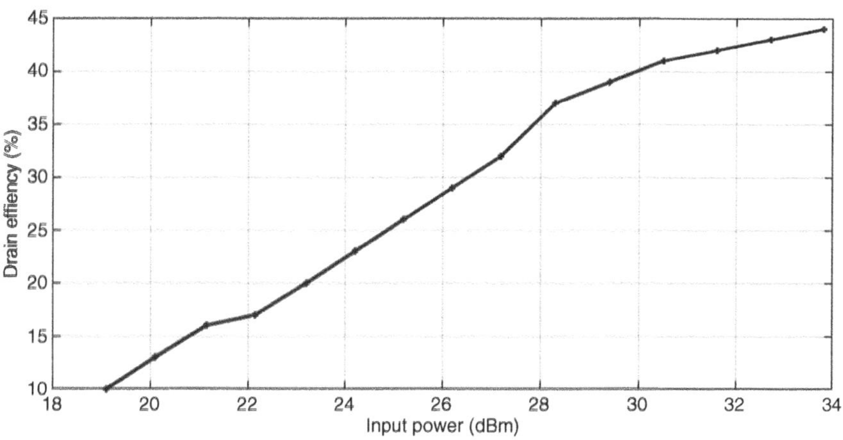

Figure 11.27 Dependence of the efficiency of the collector (drain) on the excitation level.

References

1 Kasal, M. (2013). Microwave solid state power amplifier technology. In: *13th Conference on Microwave Techniques COMITE*: 173–176. Pardubice, CR: IEEE. https://doi.org/10.1109/COMITE.2013.6545064.

2 Raab, F.H., Asbeck, P., Cripps, S.C. et al. (2002). Power amplifiers and transmitters for RF and microwave. *IEEE Transactions on Microwave Theory and Techniques* 50(3): 814–826. https://doi.org/10.1109/22.989965.

3 Cripps, S.C. (2002). *Advanced Techniques in RF Power Amplifier Design.* Norwood, MA: Artech House. ISBN 1-58053-282-9.

4 Kasal, M. (1984). Multi-nuclear measurement channel of the NMR spectrometer. Dissertation thesis. (in Czech) Brno University of Technology.

5 Keysight Technologies. (2020). *Creating an X-Parameter Model.* https://www.keysight.com/zz/en/assets/3120-1211/technical-overviews/Creating-an-X-Parameter-Model.pdf (accessed 7 August 2024).

6 Qorvo. (2019). *GaN Device Channel Temperature, Thermal Resistance and Reliability Estimates.* https://www.qorvo.com/design-hub/videos/understanding-gan-thermal-analysis (accessed 7 August 2024).

7 Triquint/Qorvo. (2016). *TGF2023-05 25 Watt Discrete Power GaN on SiC HEMT (Datasheet).* https://www.qorvo.com/products/p/TGF2023-2-05 (accessed 7 August 2024).

8 Grebenikov, A. (2005). *RF and Microwave Power Amplifier Design.* London: McGraw-Hill. ISBN 0-07-144493-9.

9 Pedro, J.C., and Maas, S.A. (2005). A comparative overview of microwave and wireless power-amplifier behavioral modeling approaches. *IEEE Transactions on Microwave Theory and Techniques* 53(4): 1150–1163. https://doi.org/10.1109/TMTT.2005.845723.

10 Mimix. (2007). *XP1006-FA 8.5-11.0 GHz GaAs Power Amplifier, Flange, 10 pin (Datasheet).* https://www.alldatasheet.com/datasheet-pdf/download/224102/MIMIX/XP1006-FA.html (accessed 7 August 2024).

11 Cree. (2016). *Datasheet of CNPA801B025F.* https://www.micro-semiconductor.com/datasheet/00-CMPA801B025F.pdf (accessed 7 August 2024).

12 Triquint/Qorvo. (2013). TGA2312-FL. *Datasheet.* https://www.qorvo.com/products/amplifiers/high-frequency-amplifiers (accessed 7 August 2024).

13 Suckling, C.H. (2015). A 50 W GaN solid state power amplifier for 10 GHz. *DUBUS Magazine* 1: 178–194. ISSN 1438-3705.

12

Communication by Lunar Surface Reflection

The use of the surface of the Moon for communication purposes represents a challenge in the area of amateur radio broadcasting at the limit of the physical possibilities of current radio technology. The development follows two directions: improvements of microwave technologies and investigations of new methods in the field of signal processing. Recently developed methods of digital communication will be presented, which can eliminate most negative effects of Earth-Moon-Earth (EME) communication. In parallel, microwave technologies for this communication will be described. Attention will be given to the 10 GHz band, parameter settings, and calibrations, including examples of implementations and experimental results.

12.1 Lunar Surface as a Reflecting Surface

In general, the lunar surface is an imperfect reflective surface due to its round shape and surface roughness. The reflection coefficient is reported to be about 6.5%, i.e. 0.065 [1]. Since the diameter of the Moon is 3 500 km, the effective reflecting surface is

$s = (3.5\times10^6/2)^2\cdot\pi\cdot0.065 = 6.25\times10^{11}\ \text{m}^2$. From the radar equation (5.13) and [1],

$$P_\text{r} = \frac{sP_\text{t}G_\text{t}G_\text{r}\lambda^2}{(4\pi)^3D^4}. \tag{12.1}$$

Here, P_t is the power of the transmitter, P_r the power at the receiver input, λ denotes the wavelength, G_t and G_r are the gains of antennas, and D is the distance of the Moon. For a wavelength of 3 cm, the attenuation of the EME path is 287.5 dB for the Moon in perigee ($D = 356.4\times10^6$ m) and 289.8 dB for the Moon in apogee ($D = 406.7\times10^6$ m). These values are approximate since some assumptions are not met (antenna beams do not overlap the Moon several times, or the reflective surface does not consist of many individual mirrors, etc.). The delay of the signal

Reception of Weak Radio Signals from Space, First Edition. Miroslav Kasal.

when traveling to the Moon and back is 2.4–2.7 seconds, which allows to receive and measure own reflections.

The average angular width of the Moon is 0.5°. For a parabolic antenna with a diameter of 4 m, the main lobe of the radiation pattern is just filled out by the Moon at a wavelength of 3 cm (see Table 12.4). With a larger antenna diameter, only a part of the lunar surface is used for reflection. Conversely, smaller antennas illuminate the vicinity of the lunar disk as well.

12.1.1 Reflected Signal Spectrum Broadening

At long wavelengths (2 m), a relatively small area around the center of the lunar disk contributes to the reflection. With increasing frequency, the region widens. And at a wavelength of 3 cm, more or less the entire area of the lunar disk contributes. The received signal consists of several components with different delays (and different phases) like after the reflection from a large number of individual mirrors.

However, the Moon swings (slightly oscillates) when traveling around the Earth. This movement is called libration. Although the Moon shows the near part (the so-called bound rotation), 59% of the Moon surface can be observed from Earth thanks to the libration. Four components of these oscillations can be distinguished. Each oscillation has a different cause: longitudinal, latitudinal, diurnal, and physical. Together, the oscillations cause the Moon to librate. Due to this relative motion, individual components of the reflected signal have different Doppler shifts and cause a wider or narrower frequency spread (see Fig. 12.1). This spectral broadening fundamentally influences the ability to decode the reflected

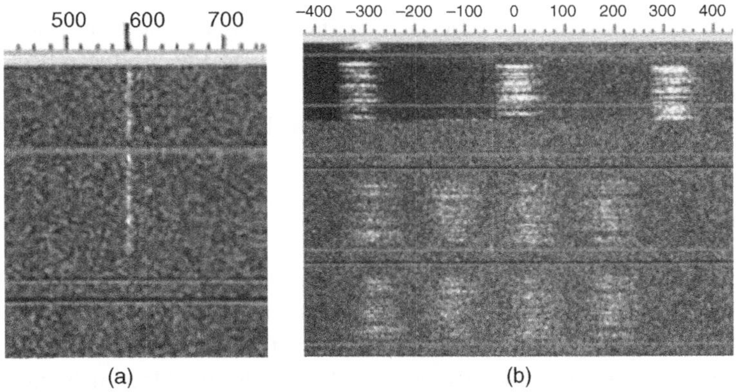

Figure 12.1 Spectrogram of harmonic signals reflected from the lunar surface in the 10 GHz band: (a) At extremely small spread. *Source: radio.feec.vutbr.cz*; (b) JT4 signals at normal spread. The upper scale is in Hz. *Source: radio.feec.vutbr.cz.*

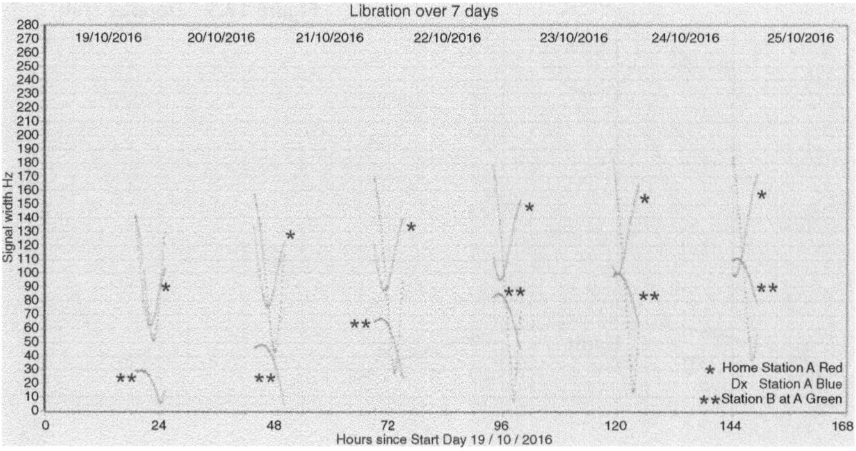

Figure 12.2 Prediction of signal spread within one week. Station A is located in the Czech Republic and station B in Japan. *Source: McArthur* [2]/*with permission of J.F.Drew.*

signals. Therefore, it is important to predict this broadening. Libration varies not only during the orbit around the Earth, but between the moonrise and moonset as well. In general, the signal spread caused by libration is smaller after the moonrise and before the moonset, while the largest values are achieved when the moon is at the peak. Therefore, the signal spread differs for observers at different locations on the Earth at a given time. Communicating stations are therefore looking for a compromise time when the spread of signals caused by libration is the most favorable for both of them. Astronomers have mastered this problem, and their algorithms can be used to calculate the libration (see Fig. 12.2, [2]).

12.1.2 Doppler Shift

So far, relatively small oscillations of the lunar surface relative to the observer were considered, which resulted in a spread of the signal due to libration. However, the lunar surface and an observer on the Earth are in a higher relative motion given by Earth's rotation and Moon's orbital motion. The resulting Doppler shift reaches ± 25 kHz in the 10 GHz band (see Fig. 12.3). At low SNR ratios, the Doppler shift has to be precisely compensated since the signal can be detected with a very narrow filter only. In these cases, the receiver has to be tuned to the correct frequency, expecting that the signal appears on the waterfall of the spectrogram. Searching for a signal by retuning the receiver is out of the question.

The mutual Doppler shift:

$$\vartheta_1 = f\frac{v_1}{c},\ \ \vartheta_2 = f\frac{v_2}{c} \quad \Rightarrow \quad \text{Mutual} \quad \vartheta = \vartheta_1 + \vartheta_2$$

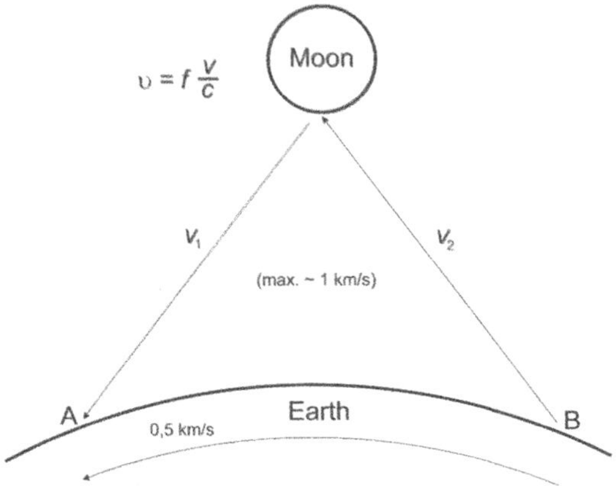

Figure 12.3 Doppler shift.

$$\text{Echo B} \Rightarrow \text{Mutual} - \frac{\text{Echo A}}{2}$$
$$\text{Full Dopp} \Rightarrow F_{\text{TX}} - \vartheta \quad \text{and} \quad F_{\text{TX}} + \vartheta$$
$$\text{CFOM} \Rightarrow F_{\text{TX}} - \frac{\vartheta}{2} \quad \text{and} \quad F_{\text{TX}} + \frac{\vartheta}{2}.$$

Full Dopp (full compensation by a single corresponding station) and CFOM (constant frequency on the Moon) are the compensation methods used.

12.1.3 Polarization of Signals

When communicating by the reflection from the lunar surface, the wave passes through the ionosphere twice. In lower frequency bands, Faraday rotation plays a role. The rotation is inversely proportional to the square of the frequency [3]. In lower microwave bands (1 296 MHz and above), the use of circular polarization is advantageous. Nevertheless, reflection changes the sense of rotation. According to the convention, the station has to be able to transmit RHCP and receive LHCP. At 10 GHz band, Faraday rotation is already negligible, and linear polarization can be used. However, the circular polarization solves the problem of the spatial angle as well. From the viewpoint of the Moon, the horizontal direction or the vertical one are not oriented identically at different places on Earth's surface, e.g. with linear polarization (LP) in the horizontal plane of the antenna in the US, the reflected signal appears vertically polarized in Europe (see Fig. 12.4). The convention for linear polarization defines that stations in Europe and Australia operate with vertical polarization and stations in the US and Japan operate with horizontal polarization. Thus, the solid angle is compensated, approximately at least. The 45° LP polarization mismatch means a loss of 3 dB. There is an effort

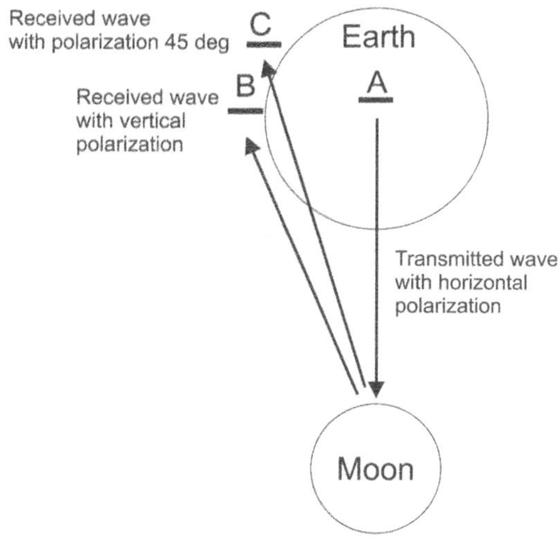

Figure 12.4 Illustration of a solid angle.

to switch to circular polarization (CP) in this band gradually. Nevertheless, the production of a high-quality feed with an axial ratio close to one is not easy [4]. Some leading experimenters [5, 6] performed a series of measurements of signal "depolarization" by reflection from the lunar surface. As shown in [6], CP is more "damaged" by reflection from the lunar surface than LP since the measurement of the self-echo of the configuration CP-CP and LP-LP comes out by 1.6 dB in favor of LP. A rotary feed with LP seems to be the best solution so far.

12.2 Digital Communication

Keying the carrier with Morse code (denoted CW) is the standard method of communication among radio amateurs using the reflection from the lunar surface. In the 3 cm band, the detected acoustic signal appears as a keyed noise against the background noise of the receiving system. Receiving such a message is quite difficult, and the difficulties raise with decreasing signal-to-noise ratio (SNR). Only in rare cases when the spread of the signal is very small the acoustic signal has the character of a tone (see Fig. 12.1a). Therefore, CW is mainly used by larger stations with large parabolic mirrors (few meters). The weak-signal radio communication group around Joseph H. Taylor, K1JT (the 1993 Nobel Prize winner for the discovery of the binary pulsar) developed communication protocols and implemented them in programs called WSJT (weak signal communications Joe Taylor) [7, 8]. E.g. JT65B used in VHF and UHF bands enables digital communication with a ratio of up to −28 dB below the noise level of the receiving system with bandwidth of 2 500 Hz. Considering that CW requires the minimum bandwidth of 100 Hz and SNR ≥ 1, the mentioned digital mode is 10–14 dB more sensitive.

This fact opens up new possibilities in communication by lunar surface reflections since the demands (on antenna dimensions especially) are significantly smaller. On the other hand, new demands appear—the absolute frequency accuracy and stability, the Doppler shift accurate compensation during reception and transmission, the precise time synchronization, etc.

12.2.1 JT65

At lower frequencies, the FSK-based JT65 with 65 tones, constant amplitude, and no phase discontinuities is the basic communication mode. The equidistant tone spacing can be 2.69, 5.38, or 10.77 Hz, which determines the total width of the signal as well. The mode description is supplemented with letters A, B, or C, accordingly. The coding is adapted exactly to the transmitted message consisting of a callsign and a grid locator or report. Each message is source-coded into six twelve-bit symbols first. FEC (forward error correction) follows and extends the message by an additional 306 redundant bits, so the redundancy ratio is 378/72 = 5.25. During the one-minute transmission interval, the message is repeated five times. This shows how robustly the message is encoded. Then, the so-called interleaving follows: symbols are row-by-row rewritten into the 7×9 matrix, and scrambled binary symbols are converted to Gray code. That way, a lower sensitivity of JT65 to frequency instability is achieved. Next to standard messages, the abbreviations RO, RRR, and 73 are most commonly used. Messages are transmitted as pairs of tones with a given spacing.

JT65 uses one-minute T/R sequences, and requires precise time synchronization between the transmitter and the receiver. Such a synchronization cannot be provided except by the actual synchronization information carried by the JT65 itself. A pseudorandom sync vector is therefore embedded among message bits as the lowest 65th tone. This enables accurate calibration of relative time and frequency, averaging of messages with the same content, and their subsequent decoding at a higher SNR. The sync vector is transmitted for half of the time of the transmit period and therefore is the brightest one in the waterfall (see Fig. 12.5).

In the 10 GHz band, the JT65C mode can be used only exceptionally since the signal spread is rarely smaller than the distance of tones. However, even messages with a certain spread of a signal spectrum beyond 10.77 Hz can be correctly decoded as shown in [9].

12.2.2 JT4

For microwave bands, a different system has to be obviously designed. The system should show sufficiently wide tone distance to allow the use by a usual spread as

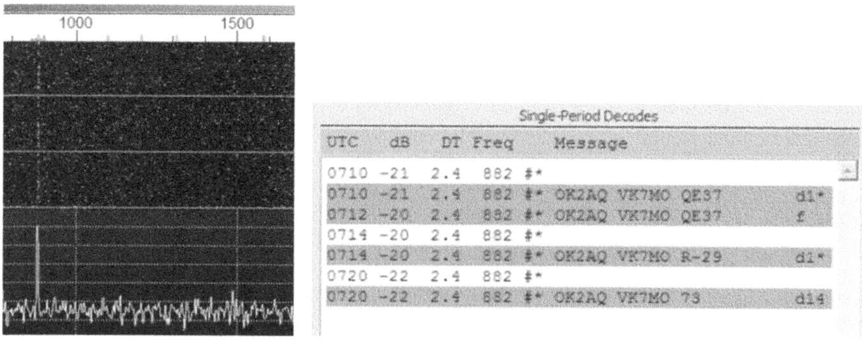

Figure 12.5 JT65C signal of the Australian station from Tasmania in the 10 GHz band at minimum spread.

well. Thus, the JT4 mode was born. In the 10 GHz band, JT4F is most commonly used, with four tones spaced at 158 Hz and a total width of 630 Hz. JT4 uses a 72-bit message structure similar to JT65. In this case, error control coding (ECC) is performed using a robust convolutional coding with length of 32 bits and code ratio of 1:2. The total length of the encoded message is then $(72+31)\times 2 = 206$ bits. In this case, the synchronization is again ensured by a binary pseudorandom sequence, but each symbol carries one information bit (the most significant one) and one synchronization bit (the least significant one). Therefore, the synchronization vector is not visible in the waterfall. A screenshot of the WSJT-X program showing the spectrogram and the communication window when connecting with the JT4F mode is depicted in Figure 12.6. Table 12.1 lists basic parameters of individual JT modes.

All the operations related to the generation of tones and sampling of received signals are performed by a PC sound card with a sampling frequency of 11 025 Hz and 16-bit resolution. A newer version of the program, WSJT-X, uses 48 000 Hz and 16 bits. Originally, the patent-protected Koetter-Vardy decoder was used for efficient RS decoding. The new "X" version already uses its own efficient FT decoder. The WSJT-X program includes other communication modes such as JT9, FT8, FT4, MSK144, FST4W, and WSPR, which are not used for EME.

12.2.3 QRA64

However, there is ongoing research of effective coding. After the period of turbo-codes, low-density parity-check (LDPC) codes were discovered, including an original decoding method. Considering that, Nico Palermo, IV3NWV, developed the QRA64 protocol [11] currently implemented by Joe Taylor in WSJT-X. The QRA64 is based on the "Q-ary repeat accumulate" code, which is

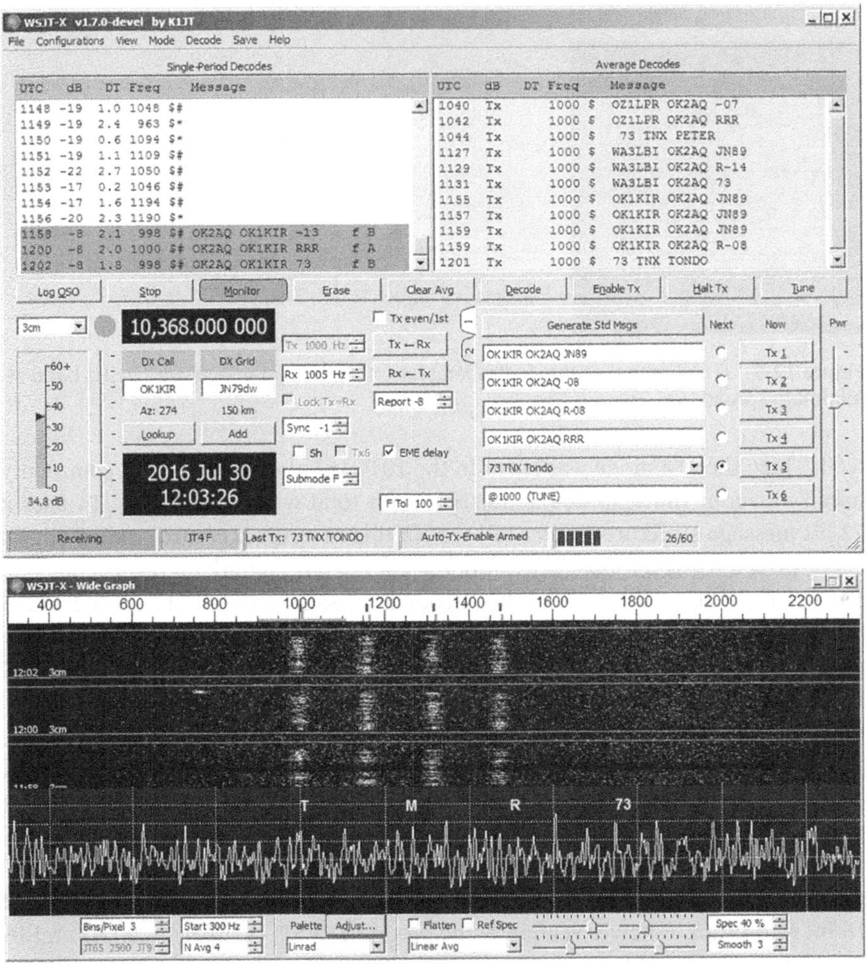

Figure 12.6 Recording of the signal and communication by the EME mode JT4F from the Všenory OK1KIR station.

obviously better than the Reed Solomon (63, 12) code used in the JT65, and gives a 1.3 dB improvement in sensitivity. QRA64 also uses a new synchronization scheme based on the Costas 7×7 vector. That way, an additional sensitivity increase of 1.9 dB is achieved. As a result, the bright sync tone is not visible in the spectrogram (see Fig. 12.7). The QRA64 mode is used similarly to the JT65. However, shortened messages are not available. Moreover, the database of call signs and locators "call3.txt" is not used as in JT65 with deep search decoding. But the sensitivity has been increased since the decoding exploits information that

Table 12.1 Overview of WSJT modes for EME [10].

Mode	T/R (s)	Mod	FEC	Nsps	Baud	df (Hz)	BW (Hz)	cps (s)	SNR (dB)
JT65A	60	65-FSK	RS(63,12)	4 096	2.69	2.69	178	46.8	−25
JT65B	60	65-FSK	RS(63,12)	4 096	2.69	5.38	355	46.8	−24
JT65C	60	65-FSK	RS(63,12)	4 096	2.69	10.77	711	46.8	−23
JT4A	60	4-FSK	$K = 32, r = 1/2$	2 520	4.375	4.375	17.5	47.1	−23
JT4B	60	4-FSK	$K = 32, r = 1/2$	2 520	4.375	8.75	35	47.1	−22
JT4C	60	4-FSK	$K = 32, r = 1/2$	2 520	4.375	17.5	70	47.1	−21
JT4D	60	4-FSK	$K = 32, r = 1/2$	2 520	4.375	39.4	158	47.1	−20
JT4E	60	4-FSK	$K = 32, r = 1/2$	2 520	4.375	78.8	315	47.1	−19
JT4F	60	4-FSK	$K = 32, r = 1/2$	2 520	4.375	158	630	47.1	−18
JT4G	60	4-FSK	$K = 32, r = 1/2$	2 520	4.375	315	1 260	47.1	−17

Nsps is the number of samples in one interval, Baud is transmission rate, df is frequency spacing between individual tones, BW is the bandwidth, cps is the total length of the transmit sequence, and SNR is the minimum decodable SNR relative to the channel bandwidth 2 500 Hz.

has been acquired during the continued connection. Averaging is not available in this mode yet but will be surely added in future versions of the WSJT-X program.

First, QRA64A was tested on short waves (30 m). Then, QRA64D was tested at EME in the 10 GHz band, especially with a large signal spread. Without problems, connection to England was established with a 127 Hz signal width. The next day, the spread was 157 Hz, and the connection failed. However, the postprocessing of the decoding showed that even these signals can be safely decoded with a different decoder setting.

12.2.4 Q65

The development of communication modes continues. Q65 is the most sensitive of all the JT mods and replaces the successful, above-described QRA64. Q65 is used for establishing a connection in case of Doppler spread of signals and short-term fading. Q65 exploits FSK 65 tones and a new low-rate Q-ary repetitive accumulation code for forward error correction (FEC). One tone at the low margin of the spectrum is used for time and frequency synchronization, and is visible on the spectral waterfall (similarly to JT65). Contrary to JT65, synchronization and decoding is efficient even in case of short-term occurrence or amplification of the signal. Q65 offers several T/R sequences that can be used according to signal strength: 15, 30, 60, 120, and 300 seconds (see Table 12.2), and efficiently uses

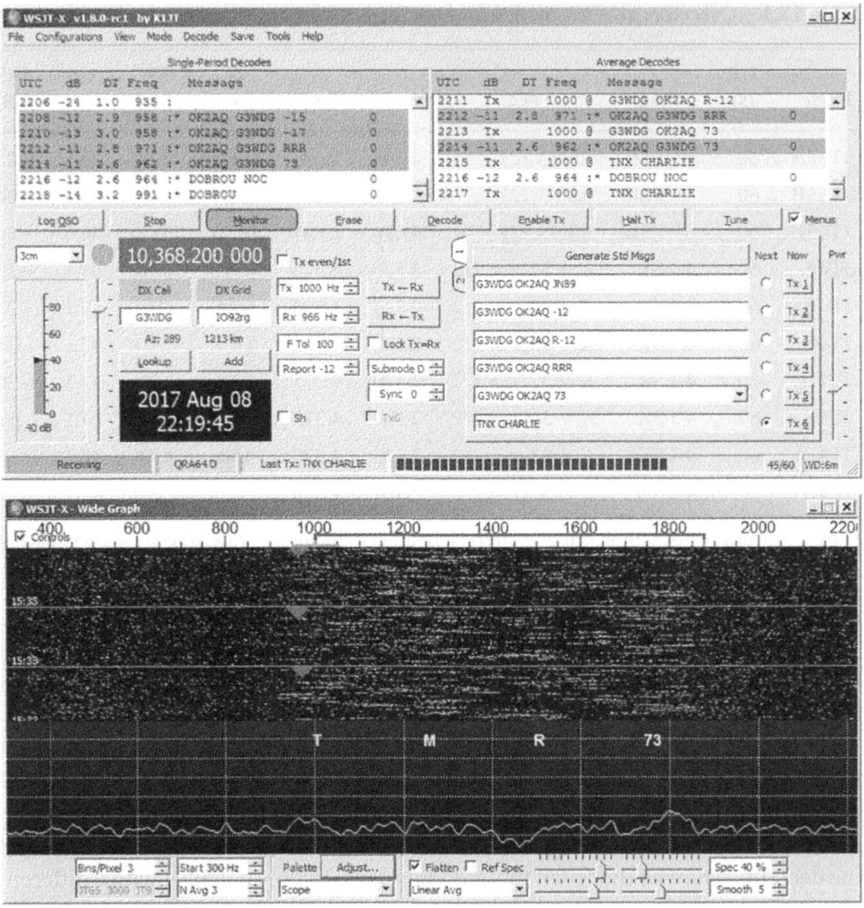

Figure 12.7 Recording of the signal and communication by the EME mode QRA64D from the England G3WDG station.

fragments of previously decoded messages during the decoding (so-called a priori decoding, AP). Sequences 30 seconds long and longer are used for EME, with 60 seconds being the default.

A specially designed block code (65, 15) with six-bit symbols is used for FEC. Two symbols are "perforated" by another code (63, 13) providing an efficient transmission ($k = 13$) of information symbols in 63 channel symbols. The "perforated" symbols contain a 12-bit cyclic redundancy check (CRC) of 13 information symbols. CRC significantly reduces false decoding. A pseudorandom sequence of 22 symbols is transmitted as "0" tone and is used for synchronization (compared to

Table 12.2 Basic parameters of the Q65 mode.

T/R period (s)	Symbol length (s)	Tone spacing (Hz)	Occupied bandwidth (Hz)	Transmission duration (s)	SNR (dB)	Max AP SNR (dB)
15	0.150	6.667	433	12.8	−22.2	−23.7
30	0.300	3.333	217	25.5	−24.8	−26.6
60	0.600	1.667	108	51.0	−27.8	−30.4
120	1.333	0.750	49	113.3	−30.8	−32.5
300	3.456	0.289	19	293.8	−33.8	−37.4

The T/R period is the period of alternating transmission and reception, symbol length is the length of the symbol, tone spacing is frequency spacing of the tones, occupied bandwidth is the bandwidth, transmission duration is the total transmission time, SNR is the achievable decodable signal-to-noise ratio at a channel width 2 500 Hz, and max AP SNR is the achievable decodable SNR at 2 500 Hz channel width using a priori information.
Source: [12] / *with permisssion of TAYLOR, J., SOMERVILLE, B., FRANKE, S., et al.*

Table 12.3 Distances of tones and spectrum width of Q65 modes.

T/R period (s)	B Spacing width (Hz)		C Spacing width (Hz)		D Spacing width (Hz)		E Spacing width (Hz)	
15	13.33	867	26.67	1 733				
30	6.67	433	13.33	867	26.67	1 733		
60	3.33	217	6.67	433	13.33	867	26.67	1 733
120	1.50	98	3.00	195	6.00	390	12.00	780
300	0.58	38	1.16	75	2.31	150	4.63	301

Source: [12]/*with permisssion of TAYLOR, J., SOMERVILLE, B., FRANKE, S., et al.*

JT65, when the synchronization signal is transmitted half the time, the synchronization signal on the spectral waterfall is therefore slightly weaker), so the total number of channel symbols at Q65 is $63 + 22 = 85$.

For individual T/R sequences, different equidistant distances of tones can be used, and also the total width of the signal spectrum according to Table 12.3.

12.3 Parameters of Communication Devices

There are different strategies to establish the EME communication. First, antennas as large as possible can be built, power as high as possible can be transmitted, and then a strong own echo can be received. Second, equipment can be minimized and all the tricks can be applied to optimize the equipment. That way, top parameters are achieved and the equipment is usable even with small antenna dimensions.

12.3.1 Antennas

For EME digital communication in the 10 GHz band, antennas with an aperture diameter higher than 1 m can already be used, as shown by results of some experimenters. However, larger antennas are optimal. The 2.4 m offset reflector seems to be suitable even when working with CW. The owners of larger reflectors have an advantage of sensitivity margin (in SSB experiments, e.g.) or of lower spread when the entire surface of the Moon is not used for reflection. However, not only the size of the reflector is important but its uniform and accurate irradiation as well (see Fig. 12.8). The over-irradiation of the reflector increases the noise temperature of the system and results in sensitivity decrease. The mechanical design of the antenna and its positioning are also very important. An antenna with a diameter of 1.8 m has a beam width of 1.1°. Hence, the manipulation step both in azimuth and elevation should not be higher than 0.2°. In case of larger reflectors, the manipulation should be proportionally more accurate (see Table 12.4). For precise operation, electronic position control in both axes is absolutely necessary.

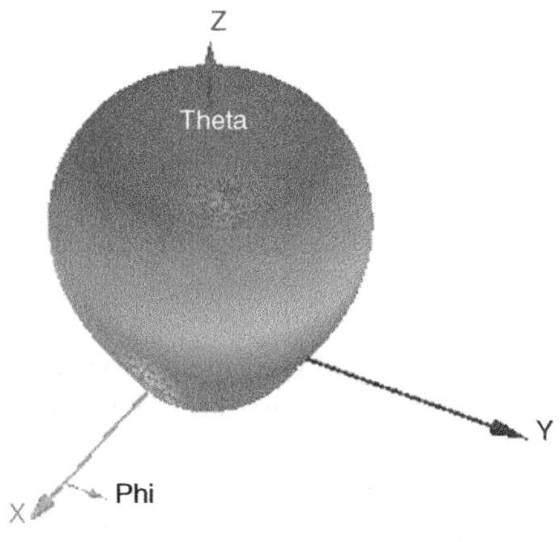

Figure 12.8 Feed design.

Table 12.4 Gain G and beam width $\theta_{3\,dB}$ of parabolic antennas with diameter D and efficiency $\eta = 0.6$ at 10 368 MHz.

D	m	1.2	1.8	2.4	3.0	4.0	5.0
G	dBi	40.1	43.6	46.1	48.0	50.5	52.5
θ_{3dB}	°	1.7	1.1	0.8	0.7	0.5	0.4

12.3.2 System Noise Temperature

From the viewpoint of sensitivity, the feed is necessary to be connected via a waveguide switch (WR90 or WR75) to a high-quality low-noise amplifier (LNA) with a waveguide input as well (see Fig. 12.9, [13]).

However, EME experiments are limited by the own noise temperature of the Moon [14] (see Fig. 12.10b, [15]). The brightness temperature of the Moon is relatively constant. At a frequency of 10 GHz, $T_{Mo} \approx 200$ K. The charts in Figure 12.10a [15] and Figure 12.10c [15] show that the noise increases for $\Delta T_{Mo} = 20$ K $(0.1T_{Mo})$ if the ratio of the antenna beam width (1.13°) to the angular width of the Moon is 2.26. If the brightness temperature of the clear sky $T_{SKY} = 5$ K, contribution from the Earth $T_G = 40$ K, and the noise temperature of the receiver $T_R = 52$ K (corresponding to noise figure 0.7 dB measured at the temperature 290 K) are considered, then the noise increase for an antenna with an aperture diameter of 1.8 m is MN = 0.81 dB. For an antenna with a diameter of 4.5 m, MN = 3.07 dB. Obviously, the smaller the antenna, the more important the low noise figure of the receiver. Nevertheless, this does not help in case of large antennas. The chart in Figure 12.11 illustrates the dependency quantitatively. Here,

$$T_S = T_{SKY} + T_G + T_{Mo} + T_R = T_e + T_R,$$

$$T_e = T_{SKY} + T_G + T_{Mo},$$

$$T_{Mo} \approx 200 \text{ K} \Rightarrow \quad T_s = T_{SKY} + T_G + T_{Mo} + T_R,$$

$$\Delta T_{Mo} \approx 20 \text{ K} \quad \Rightarrow \quad T_{sCS}^{F=0.7 \text{ dB}} = 5 + 40 + 0 + 52 = 97 \text{ K},$$

$$T_{sMo}^{F=0.7\text{dB}} = 5 + 40 + 20 + 52 = 117 \text{ K},$$

$$\frac{\Theta_{3dB}}{D_\Theta} = \frac{1.13}{0.5} = 2.26, \text{ and}$$

$$\text{MN} = 10 \log(T_{sMo}/T_{sCS}) = 10 \log(117/97) = 0.81 \text{ dB}.$$

The calculated value can be measured as well (see Fig. 12.12). The sensitivity of the receiving system can be verified by measuring the noise level of the Sun as well. But this measurement is less accurate since the Sun's radiation varies more (see also Chapter 6). On the other hand, the significantly higher values are suitable for starting antenna position calibration and other experiments (see Fig. 12.13).

12.3.3 Echo

Measuring the own echo is the easiest way to verify the ability to communicate (this is possible due to the signal delay of about 2.5 seconds). The WSJT program

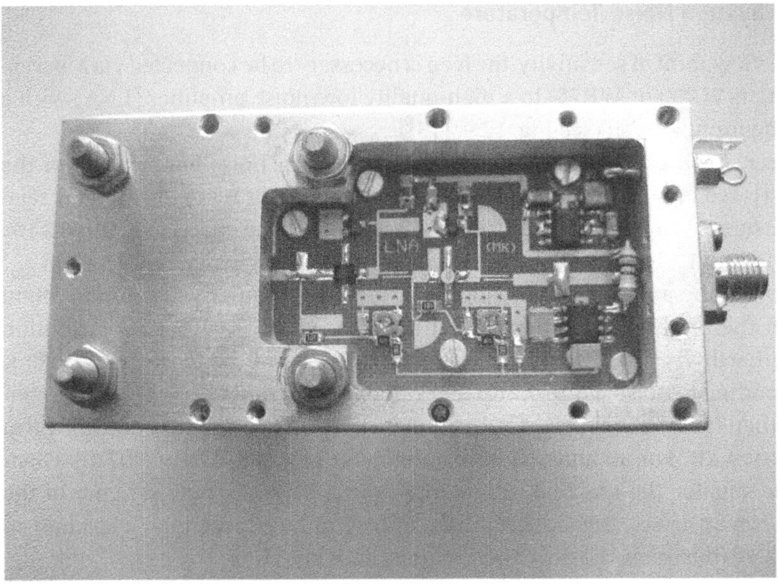

(a)

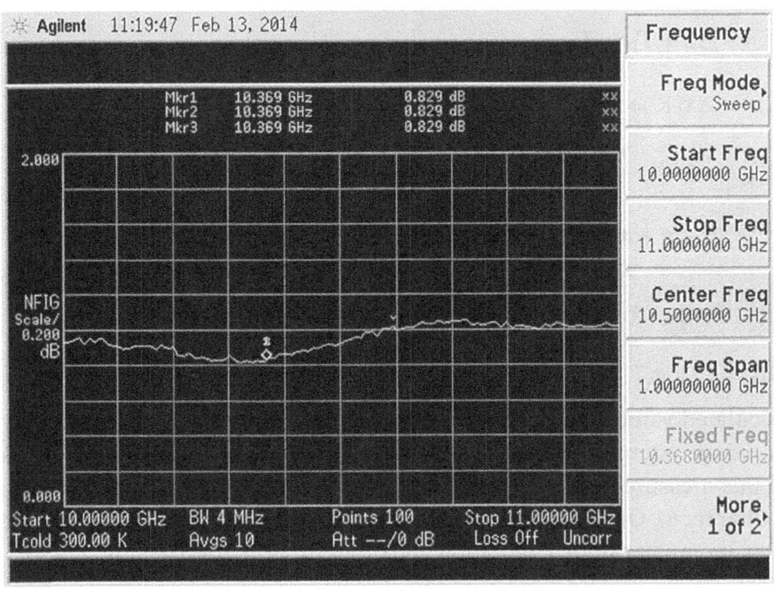

(b)

Figure 12.9 (a) The LNA with a waveguide input. *Source: radio.feec.vutbr.cz;* (b) The noise figure measured with a WR75/SMA adapter (0.08 dB).

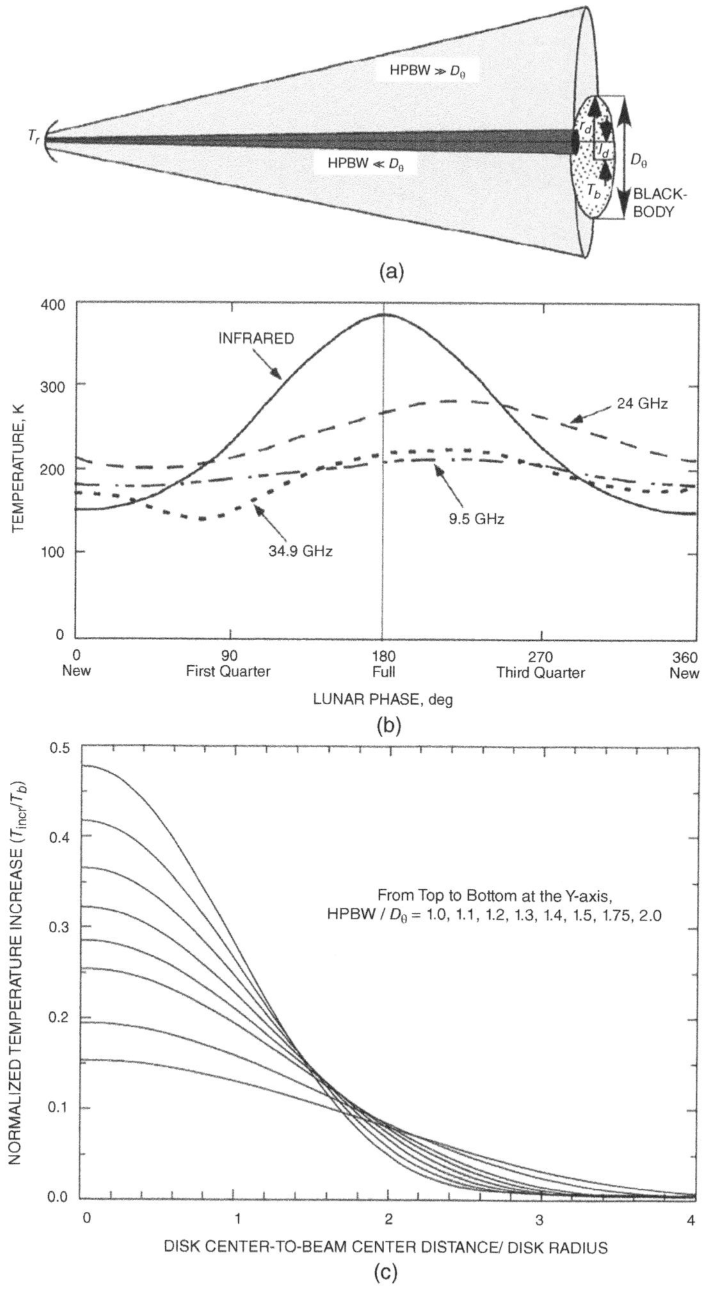

Figure 12.10 (a) The position of the antenna beam relative to the blackbody disk; (b) The brightness temperature of the Moon for different frequencies as a function of the Moon's phase; (c) The contribution of the Moon's noise temperature for different wide antenna beams and their displacement relative to the center of the blackbody. *Source: Ho et al.* [15]*/NASA/Public domain.*

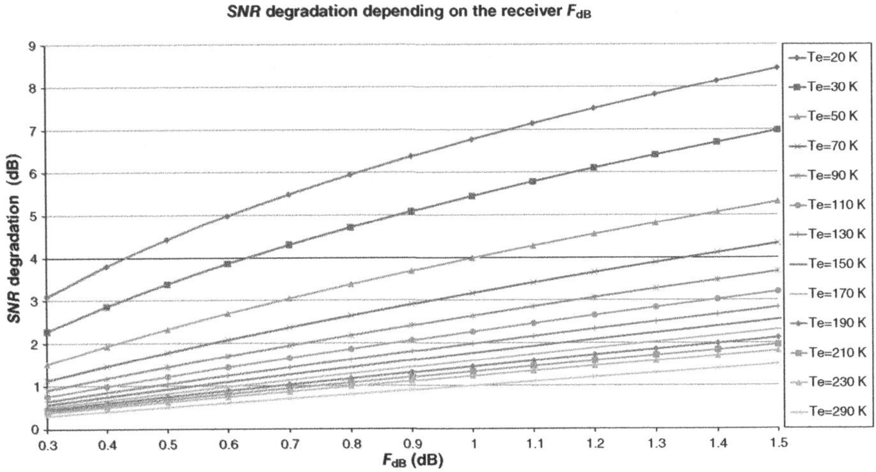

Figure 12.11 SNR degradation as a function of the noise figure of the receiver. The parameter is the equivalent noise temperature, T_e, of the whole system.

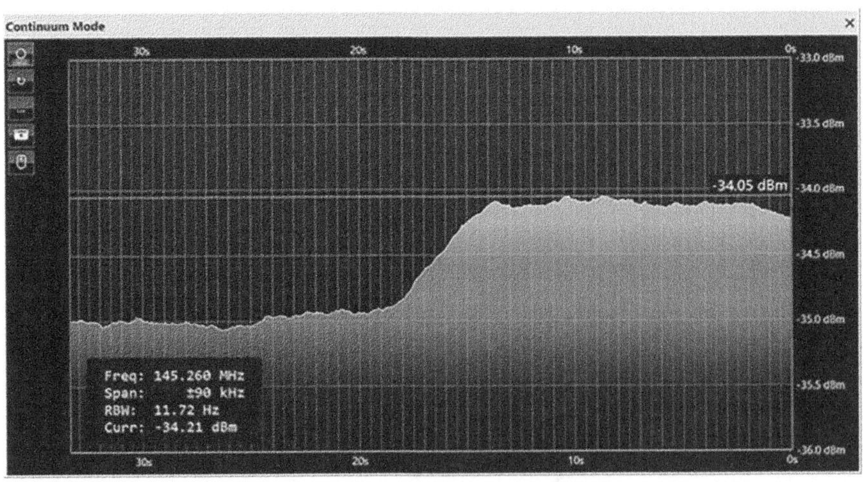

Figure 12.12 Measuring the MN/CS ratio.

already implemented this method. After starting the experiment, a 1 500 Hz tone is transmitted for 2.3 seconds, followed by a 0.2 second pause and a 2.3 second decoding interval. This sequence is regularly repeated, and the result is averaged. That way, the echo signal is usually identified after a certain number of attempts, even if weak (see Fig. 12.14). Figure 12.15 shows the echo after each sequence since the signal has a small spread. The figure also shows that the Doppler shift

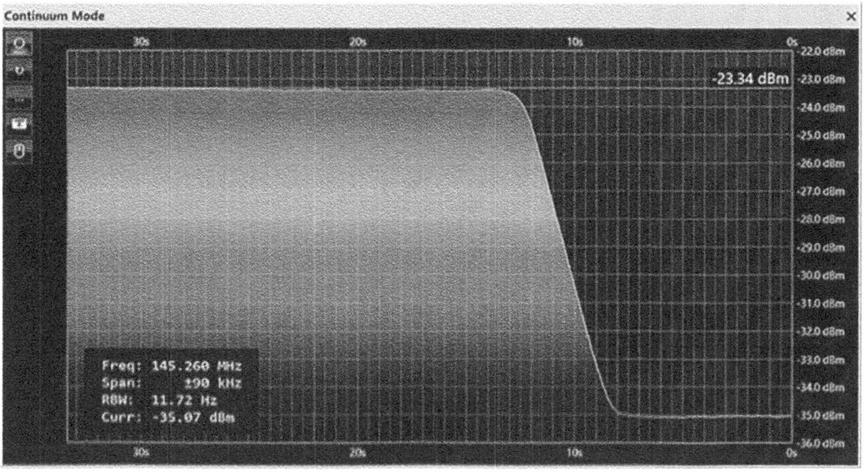

Figure 12.13 Measuring the SN/CS ratio.

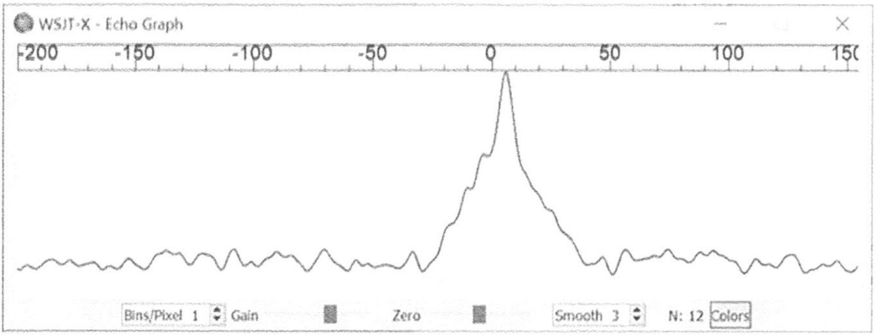

Figure 12.14 Echo measurement with averaging.

was compensated for one part of the experiment and was not compensated for another part. Naturally, the radiated power influences the echo level. In the 10 GHz band, the power of tens of watts (needed for EME experiments) is not easy to be achieved. A number of stations exploit traveling wave tubes (TWT) requiring high voltage. Since the tube is usually located in an external environment, it is difficult to ensure safety.

The 50 W solid-state power amplifier (SSPA) with a GaN-based module (see Fig. 12.16) was successfully designed and implemented. SSPA was installed together with a low-noise amplifier (LNA), a WR75 waveguide switch, and a sequencer at the feed in the focus of the reflector to minimize losses (see

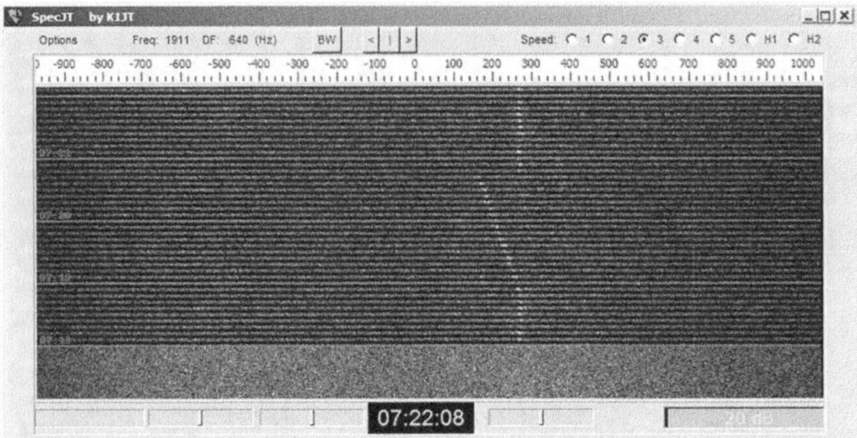

Figure 12.15 Echo measurement by sequences.

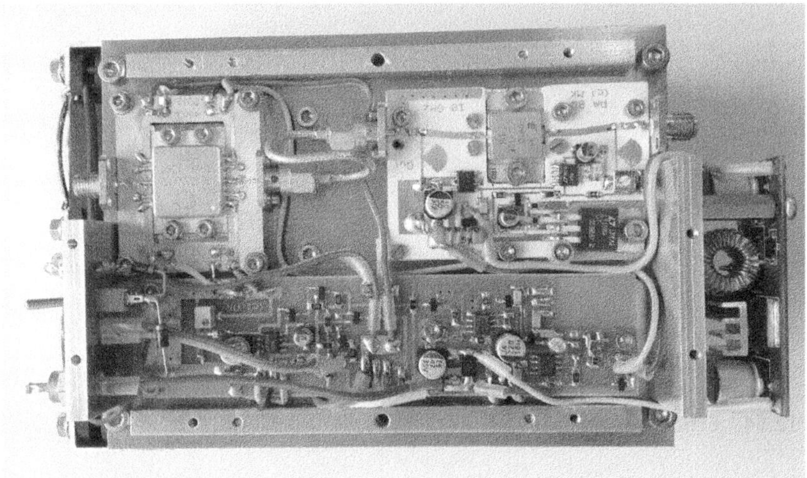

Figure 12.16 The 50 W solid-state power amplifier (SSPA), 10 GHz. *Source: radio.feec.vutbr.cz.*

Fig. 12.17). In digital modes, the transmitter operates with a duty cycle of 40%, and a corresponding cooling is required.

12.3.4 Accuracy and Stability of Frequency

Absolute frequency accuracy in the order of at least 10^{-8} is necessary since weak signals cannot be practically searched for by fine-tuning, but can only be identified on the waterfall of the spectrogram with previous fine-tuning. Local oscillators of

Figure 12.17 SSPA, LNA, and WR75 switch in focus.

the transverter, but preferably also the intermediate frequency transceiver, have to be therefore attached to the atomic reference frequency standard. Today, disciplined oscillators controlled by GPS signals or a rubidium normal are preferably used. Those oscillators guarantee frequency stability. Since accurate time synchronization of the computer is required for digital modes, the 1PPS output of the GPS module can be advantageously used for this purpose.

12.3.5 Doppler Shift Compensation

Calculation of the Doppler frequency shift in the 10 GHz band has to be very accurate for the same reason as explained in the previous paragraph. The WSJT-X program was significantly improved in this regard. Calculation based on accurate algorithms developed by the Jet Propulsion Laboratory (JPL) is saved into a text file at second intervals and displayed in the astronomical data window. Since the automatic Doppler shift compensation is a fundamental issue in digital modes, the program allows to perform this compensation using the CAT bus of the intermediate frequency transceiver. Several Doppler shift compensation methods are used, most commonly FullDopp and CFOM.

In our case, two problems were solved. Next to the accurate compensation of the Doppler shift in both periods, an IF transceiver operating at 148 MHz was hanged on a rubidium oscillator (an intervention into the device was requested). The solution was found in the construction of a 148–18 MHz transverter with a local oscillator consisting of a DDFS (direct digital frequency synthesizer) suspended on an

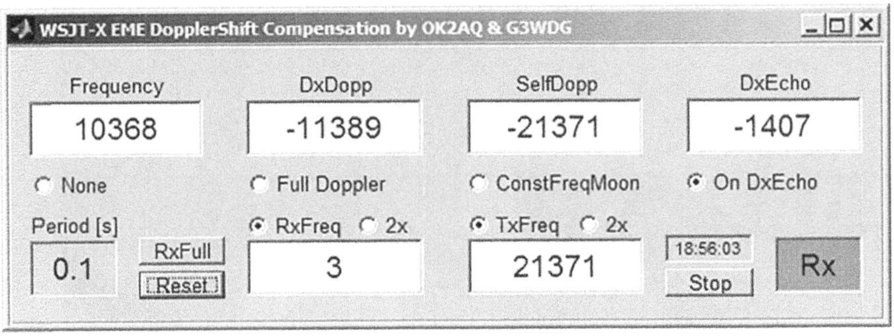

Figure 12.18 User's interface of the Dopp program.

atomic oscillator [16]. The transceiver itself operates at the frequency of 18 MHz. Here, the frequency uncertainty is low tens of Hz maximum when using a thermally compensated crystal oscillator (TCXO).

The DDFS of the transverter is controlled by the Dopp program written in MATLAB and compiled by the 32-bit MCR (Matlab Compiler Runtime). The program reads accurate data from the WSJT-X text file, calculates the current values for various methods of Doppler shift compensation, and controls the synthesizer (see Fig. 12.18).

12.4 Conclusion

This chapter summarized the basics of current EME communication in the 10 GHz band, especially the digital modes. Operating with weak signals reflected from the surface of the Moon is fascinating and incredibly informative. Especially with small systems, any mistake or even just a bad estimate, can cause a negative result. That's the beauty of it—the result is not given in advance and every time has to be "earned" by highly qualified work in several fields.

References

1 Fehrenbach, J. (2015). *What is Different on 10 GHz EME?* https://pa0ehg.com/whatis.htm (accessed 7 August 2024).

2 McArthur, D. (2019). *Libration Calculator*. https://www.vk5dj.com/doug.html (accessed 7 August 2024).

3 Kasal, M. (2006). Influence of nature on satellite communication (in Czech). *Propagation of Electromagnetic Waves*: 15–18. Prague: Czech Electrotechnical Society. ISBN 80-02-01865-6.

4 Galuscak, R., Modderman, B., Masek, V. et al. (2014). Circular polarization feed for space communication on the 3 cm band. *DUBUS Magazine* 3, 4. ISSN 1438-3705.
5 Jelinek, T., Masek, V., Karel, J. (2006). *OK1KIR Report from 10 GHz Polarization Test Held on April 2, 2006*. http://www.sm5bsz.com/linuxdsp/usage/pol10ghz.htm (accessed 7 August 2024).
6 Suckling, C. (2016). *Experiences with CP on 10 GHz*. http://www.eme2016.org/wp-content/uploads/2016/08/EME-2016-G3WDG-Presentation.pdf (accessed 7 August 2024)
7 Taylor, J. (2023). *WSJT Home Page*. wsjt.sourceforge.io/ (accessed 7 August 2024).
8 Taylor, J. (2016). WSJT-X new codes, modes and tools for weak-signal communication. *EME Conference*. Venice, Italy. https://wsjt.sourceforge.io/K1JT_EME_2016_Venice.pdf (accessed 7 August 2024).
9 Moncur, R. and Taylor, J. (2013). *Small Station EME at 10 and 24 GHz: GPS Locking, Doppler Correction, and JT4*. https://wsjt.sourceforge.io/small_station_eme.pdf (accessed 7 August 2024).
10 Taylor, J. (2016). *WSJT-X User Guide*. https://wsjt.sourceforge.io/wsjtx-doc/wsjtx-main-2.6.1.html (accessed 7 August 2024).
11 Palermo, N. (2016). Q-ary repeat-accumulate codes for weak signals communications. *XVII EME Conference*. Venice, Italy, 2016. http://www.eme2016.org/wp-content/uploads/2016/08/EME-2016-IV3NWV-Presentation.pdf (accessed 7 August 2024).
12 Taylor, J., Somerville, B., Franke, S. et al. (2021). *Quick-Start Guide to Q65*. https://wsjt.sourceforge.io/Q65_Quick_Start.pdf (accessed 7 August 2024).
13 Kasal, M. (2014). *Low Noise Amplifiers for 10 GHz and Their Measurement* (in Czech). https://www.radio.feec.vutbr.cz/esl/files/EME/Doc/LNA%2010GHz.pdf (accessed 7 August 2024).
14 Kasal, M. (2015). *EME on Microwaves with New Technologies* (in Czech). https://www.radio.feec.vutbr.cz/esl/files/EME/Doc/Gajow2015.pdf (accessed 7 August 2024).
15 Ho, C., Kantak, A., Slobin, S. et al. (2007). Link analysis of a telecommunication system on Earth, in geostationary orbit, and at the Moon: Atmospheric attenuation and noise temperature effects. *IPN Progress Report*: 42–168. https://citeseerx.ist.psu.edu/document?repid=rep1&type=pdf&doi=94f688d1637a49b5c6d69723e23d59ae119ac01e (accessed 7 August 2024).
16 Kasal, M. and Suckling, C. (2016). A new method for microwave EME Doppler shift compensation. *DUBUS Magazine* 1. ISSN 1438-3705.

13

Radio Communication with Interplanetary Spacecrafts

Radio communication in interplanetary and even in interstellar space is among the top applications of today's radio technology. Communication with the ground segment of the DSN (Deep Space Network) runs in microwave bands, and is managed by the Jet Propulsion Laboratory (JPL) in Pasadena. The DSN will be briefly described, and the most interesting projects from the viewpoint of radio communication (Voyager 1 and 2, Cassini-Huygens, New Horizons, MRO-Curiosity) will be characterized, including spacecrafts at Lagrange points.

13.1 Terrestrial Segment for Deep Space Communication

The ground segment of the DSN consists of three similarly equipped Deep Space Communication Complex (DSCC) workplaces with several Deep Space Stations (DSS). DSCCs are located in Goldstone (USA) with DSS14, 24, 25, 26, near Madrid (Spain) with DSS63, 65, 54, 55, and near Canberra (Australia) with DSS43, 34, 35, and 36. Their location is determined by the requirement to cover "all" the directions (see Fig. 13.1, [1]) and to receive extremely weak signals coming from space without disturbances. DSS differs in the number and the own antenna. The first DSS in each DSCC has an antenna with an aperture diameter of 70 m. The next three to five DSSs have antennas with an aperture diameter of 34 m, which gradually replace older antennas with aperture diameter of 26 m.

The network is managed by the Jet Propulsion Laboratory in Pasadena. This laboratory also implements most projects of spacecrafts for distant objects in space. The web page [2] shows the current status: which DSS is currently communicating and which space object is talking to (see Fig. 13.2).

Communication usually runs in the L, S, X bands, and also in the Ka band in recent years.

Reception of Weak Radio Signals from Space, First Edition. Miroslav Kasal.

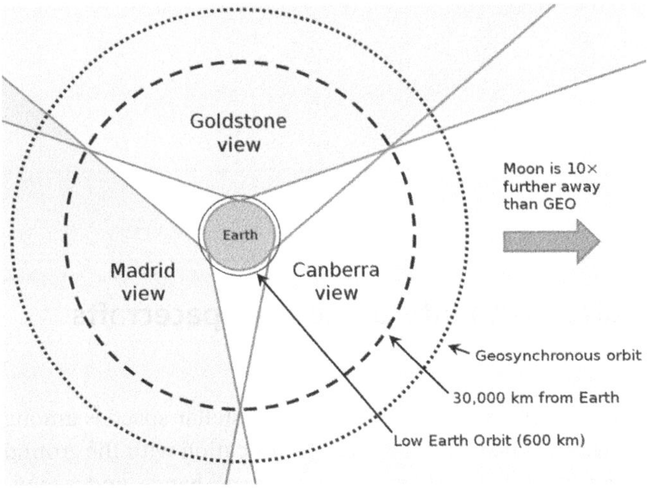

Figure 13.1 Space coverage by DSN stations. *Source: Wikipedia* [1]/*NASA Deep Space Network. (2023)/Wikipedia/CC BY 4.0.*

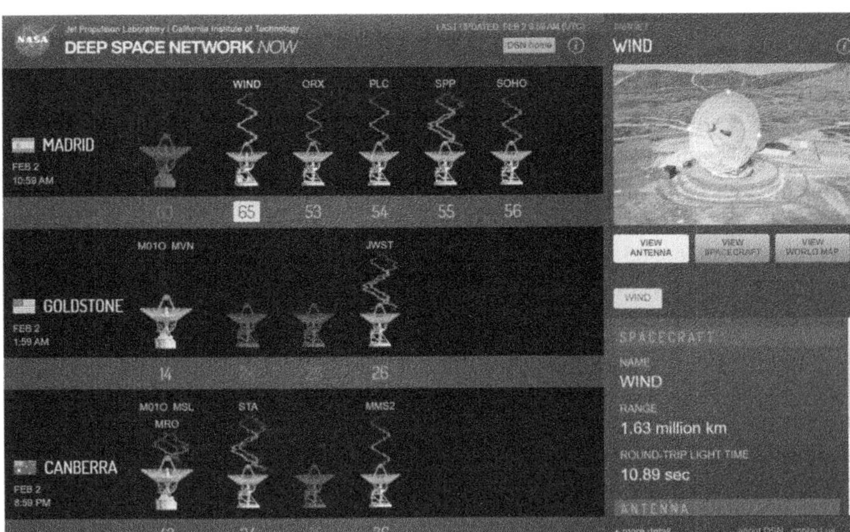

Figure 13.2 DSN network activity. *Source:* [2]. *NASA. (2023). Deep Space Network Now.* https://eyes.nasa.gov/dsn/dsn.html *(accessed 23 April 2025). NASA/Public domain/* https://eyes.nasa.gov/dsn/dsn.html *(accessed 29 April 2025).*

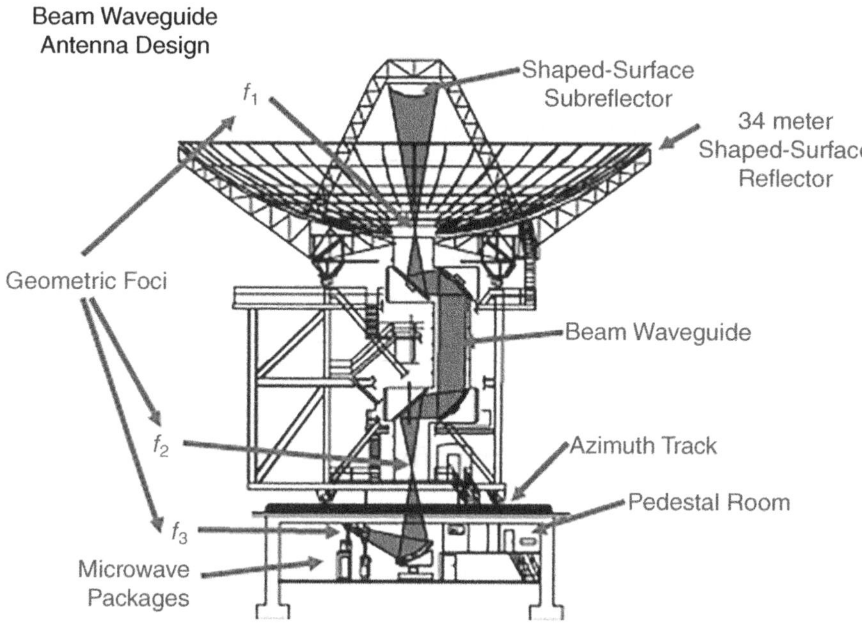

Figure 13.3 DSS antenna of the type BWG. *Source: Wikipedia* [3]*/Beam Waveguide Antenna (2011)/Wikipedia/CC By 4.0.*

DSS14, DSS63, and DSS43 with the 70 m aperture have 20 kW transmitters for the S band and the X band for uplink. Only the DSS43 has an additional 400 kW transmitter for the S band. The S/X bands are separated by a dichroic mirror and diplexers. RHCP and LHCP circular polarizations are used both for uplink and downlink. Receivers are equipped with high-quality low-noise amplifiers (LNA). Until 2010, LNA with masers were used. Now, LNA are based on HEMT amplifiers that are cooled in cryostats for the higher X, Ku, and Ka bands.

There are two types of DSS with 34 m antennas: beam waveguide (BWG, see Fig. 13.3) and high efficiency (HEF). The BWG – DSS35 are the latest in development. Since DSS are designed very flexibly, the optimal configuration can be selected for each application, e.g. if the DSS currently operates in one band, the diplexer can be disabled to achieve a lower noise temperature [4]. In addition to the DSN, the DSS can be used for radio astronomy purposes. And conversely, other radio telescopes can be included in the DSN network for a specific purpose, if necessary. The fundamental parameters of DSS stations are given in Table 13.1.

DSS stations communicate with space objects in three basic modes:

1) Radiometric data – Doppler and ranging
2) S/C commanding – stream, file, command delivery
3) Telemetry demodulation and decoding

Table 13.1 Typical DSS parameters.

DSS	Band	AntGain [dBi]	HPBW [deg]	Power [kW]	EIRP [dBm]	Noise temp [K]	FreqUpl [MHz]	FreqDwl [MHz]
34 m BWG	S	56.3/56.8	0.263	20	128.7	26.1	2 110–2 118	2 200–2 300
	X	67.1/68.2	0.077	20	139.6	20.2	7 149–7 188	8 400–8 500
	Ka	79.5/78.9	0.016	0.8	138.2	27.9	34 315–34 415	31 800–32 300
70 m	S	63.0/63.6	0.128	20/400	135.6/148.7	10.5	2 110–2 118	2 200–2 300
	X	73.2/74.6	0.038	20	145.8	20.2	7 149–7 188	8 400–8 500

Source: [5]/*JET PROPULSION LABORATORY/with permission of NASA.*

13.2 Interplanetary Spacecrafts

13.2.1 Voyager

Around 1970, researchers found out that thanks to the position of the outer planets of the Solar System, a single spacecraft would be able to fly around Jupiter, Saturn, Uranus, and Neptune in 1977. A similar situation will occur in 176 years. Within three years, Voyager 1 and Voyager 2 were prepared on the Mariner platform to launch (see Fig. 13.4). Five years after the launch of Pioneer 10 and 11, Voyager 2 was launched on August 20, 1977, followed by Voyager 1 16 days later (see Fig. 13.5). Undoubtedly, this was one of the most successful space missions with implications for the present and perhaps even the next decade. These spacecrafts are very interesting from the viewpoint of radio communication as well since both of them have already left the Solar System [6]. When writing this chapter, Voyager 2 was more than 17.5 billion kilometers (117 AU) away from the Earth: this distance is related to the attenuation 304.2 dB in the S band and 315.7 dB in the X band. The spacecraft response time is 1.35 days (1 944 minutes). At present, only DSS in Canberra is communicating with Voyager 2 since it is not visible from the northern hemisphere. Voyager 1 was directed above the ecliptic by the last maneuver and communicates with DSS in Madrid and Goldstone.

Some equipment (low-gain antenna, S band downlink) was used at the beginning of the mission only. The S band uplink frequencies are 2 114.676697 MHz (Voyager 1) and 2 113.312500 MHz (Voyager 2), respectively. The carrier can be unmodulated, modulated by a command or a ranging sequence, or both simultaneously. The receiver of the communication transponder contains a phase-locked loop (PLL); the locked VCO forms a reference for the downlink, which is coherent with the uplink. If the VCO is not locked, then the downlink carrier is derived from the ultra-stable oscillator (USO) 1×10^{-12}, which is backed up as well.

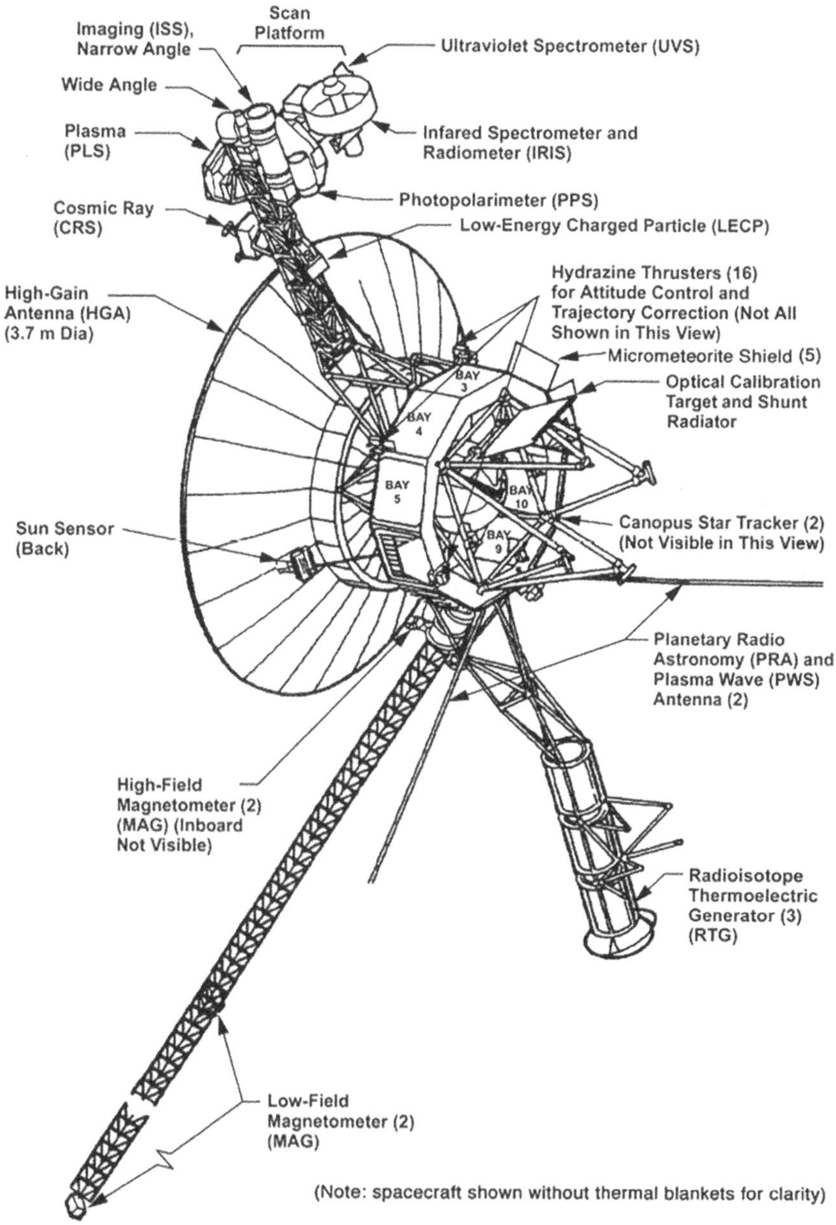

Figure 13.4 Voyager spacecraft. *Source: Ludwig and Taylor* [6]/*Ludwig, R. and Taylor, J. (2002)/Voyager Telecommunications/NASA/Public Domain.*

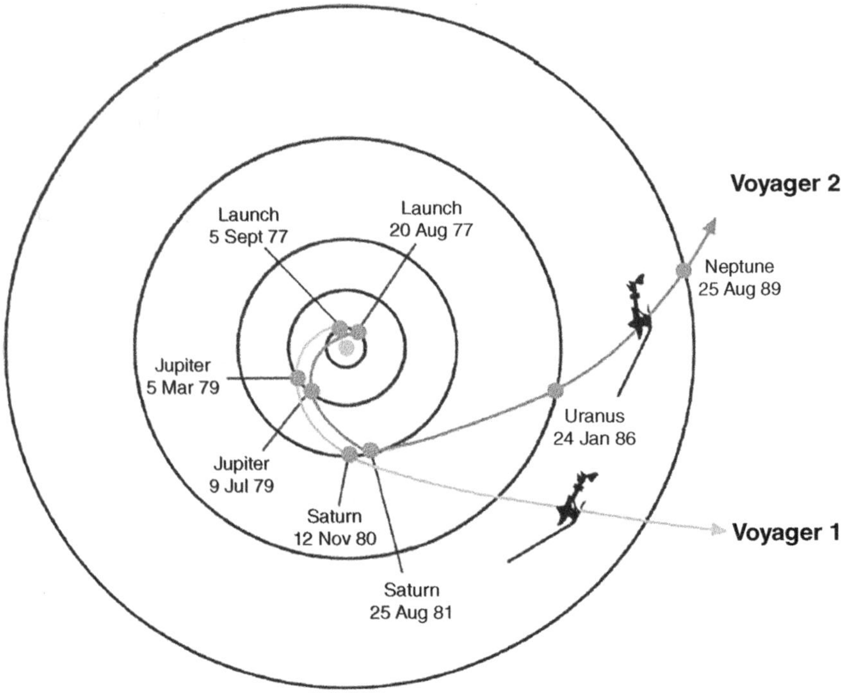

Figure 13.5 Voyager trajectories in interplanetary space. *Source: Ludwig and Taylor* [6]/ *Ludwig, R. and Taylor, J. (2002)/Voyager Telecommunications/NASA/Public Domain.*

Modulation for ranging: The standard DSN sequence for ranging is demodulated and remodulated on a downlink carrier in the S band or the X band or both. The transponder must be operated in the coherent mode.

Commands are modulated on a rectangular uplink sub-carrier of 512 Hz by BPSK 16 bps modulation with Manchester encoding.

Telemetry exploits two independent channels. The low-speed channel is modulated by uncoded NRZ BPSK 40 bps. The modulation rate of the high-speed channel varies from 10 bps to 115.2 kbps and is modulated with fixed-length coded symbols ($k = 7$, $r = 1/2$). Each BPSK modulator can modulate the 22.5 kHz or the 360 kHz sub-carrier (the second one can be used for rates higher than 7.2 kbps).

Downlink carriers: When the transponder is operating in two-way coherent mode, the ratios of downlink/uplink frequencies are 240/221 in the S band and 840/221 in the X band. In the X band, the downlink frequencies are 8 420.432097 MHz in case of Voyager 1 and 8 415.000000 MHz in case of Voyager 2. In the non-coherent mode, both frequencies are swapped. For the X band, both spacecrafts have two transmitters with TWT; the first one generates the RHCP wave and the second one the LHCP wave.

Antennas: A high-gain antenna consists of an axisymmetric parabolic reflector with a diameter of 3.7 m and a system of feeds. The gain is 36 dBi RHCP in the S band and 48 dBi in the X band (for RHCP and LHCP, alternatively). **A low-gain** antenna is installed for the S band only; its gain is 7 dBi. The low-gain antenna was used after launch immediately (before the spacecraft could orient itself). After 1980, the low-gain antenna was no longer sufficient due to the distance of the spacecrafts.

The way of developing communication parameters of the Voyager system is evident from Table 13.2 and the following overview.

The current energy balance of communication with the Voyager 2 spacecraft:

S band uplink

$$TxPower\ 20\,\text{kW} = 43\,\text{dBW},\quad G_{T34} = 56.3\,\text{dBi},\quad \text{EIRP} = 99.3\,\text{dBiW}$$

$$G_{T70} = 63.0\,\text{dBi},\quad \text{EIRP} = 106.0\,\text{dBiW}$$

$$G_R = 36\,\text{dBi},\quad T_s = 40\,\text{K} \triangleq {}^\circ 16\,\text{dB(K)},\quad G/T = 36 - 16 = 20\,\text{dBi/K}$$

$$f_s = 600\,\text{bps} \triangleq 27.8\,\text{dB(s)}$$

$$E_b/N_0 = \text{EIRP} - L_0 + G/T + 228.6 - f_s$$

$$= 99.3 - 304.2 + 20 + 228.6 - 27.8 = 15.9\,\text{dB}\ \ (34\,\text{m})$$

$$= 106.0 - 304.2 + 20 + 228.6 - 27.8 = 22.6\,\text{dB}(70\,\text{m})$$

Table 13.2 Estimation of lifetime and communication parameters over the years.

	Voyager 1	Voyager 2
Electrical input	2023	2023
Telemetry transmission:		
7 200 bps, 70/34 m HEF	1994	1998
1 400 bps, 70 m antenna	2007	2011
600 bps, 70 m antenna	2026	2030
600 bps, 34 m HEF antenna	2003	2007
160 bps, 34 m HEF antenna	2024	2029
40 bps, 34 m HEF antenna	2050	2057
Hydrazine for position control	2040	2048

Source: [6]/*Ludwig, R. and Taylor, J. (2002)/Voyager Telecommunications/NASA/ Public Domain.*

Band X downlink

$$Tx\ Power\ 18\,\text{W} = 12.5\,\text{dBW},\ G_T = 48.0\,\text{dBi},\quad \text{EIRP} = 60.5\,\text{dBiW (high power)}$$

$$G_{R34} = 68.2\,\text{dBi},\quad T_s = 20.2\,\text{K} \triangleq 13.0\,\text{dB(K)},\quad G/T = 68.2 - 13.0 = 55.2\,\text{dBi/K}$$

$$G_{R70} = 74.6\,\text{dBi},\quad T_s = 20.2\,\text{K} \triangleq 13.0\,\text{dB(K)},\quad G/T = 74.6 - 13.0 = 61.6\,\text{dBi/K}$$

$$f_s = 600\,\text{bps} \triangleq 27.8\,\text{dB(s)}$$

$$\begin{aligned} E_b/N_0 &= \text{EIRP} - L_0 + G/T + 228.6 - 27.8 - f_s \\ &= 60.5 - 315.7 + 55.2 + 228.6 - 27.8 = 0.8\,\text{dB (34 m)} \\ &= 60.5 - 315.7 + 61.6 + 228.6 - 27.8 = 7.2\,\text{dB (70 m)} \end{aligned}$$

13.2.2 Cassini-Huygens

In 1997, the Cassini spacecraft was launched in order to study Saturn in detail. The journey to Saturn took seven years using the gravity of Venus (twice), the Earth, and Jupiter in the first three years. The Cassini spacecraft carried the Huygens probe, which made a soft landing on the Titan moon, which has a thick hydrocarbon atmosphere [7]. Ultimately, the mission was very successful and substantially expanded knowledge of this part of the universe. However, the development of the mission was very dramatic. The uplink and the downlink of the Cassini spacecraft with DSN stations operated in the X band. Both links provided all three basic functions (commanding, telemetry, and radiometry). Moreover, the spacecraft was responsible to transmit data from the Huygens probe (transmitted in the S band) to the Earth. But during the flight to the planet, a fatal error in the design of the Huygens probe was revealed. The S band radio-electronic system of the descent probe did not allow the Doppler shift compensation needed to ensure the synchronization for demodulating the data that the probe was supposed to transmit during and after the landing on Titan's surface. Hence, the most interesting part of the mission was seriously endangered. After several months of efforts, a solution was found. The trajectory of the Cassini spacecraft was changed so that the relative motion between the spacecraft and the Huygens probe could be minimal during the descent and activity on the surface of Titan. Thus, the Doppler shift can be minimal as well. Because of this change, a very low Cassini fly-by of Titan (1 200 km) was omitted since the new orbit was at an altitude of 60 000 km.

The typical configuration of DSN stations is shown in Figure 13.6. When moving along the *inner* orbit, the Cassini spacecraft used low-gain antennas (LGA). When moving along the *outer* orbit, high-gain antennas (HGA) were used with the necessary precise orientation. In this case, a rate of 500 bps with a modulation deviation of 1.3 rad and a 3 dB suppressed carrier were used as a standard. The downlink for telemetry transmission had a speed of 14 220 bps with modulation deviation of 80° and ranging of 0.27 rad. Threshold values E_b/N_0, or P_r/N_0, were the same for both

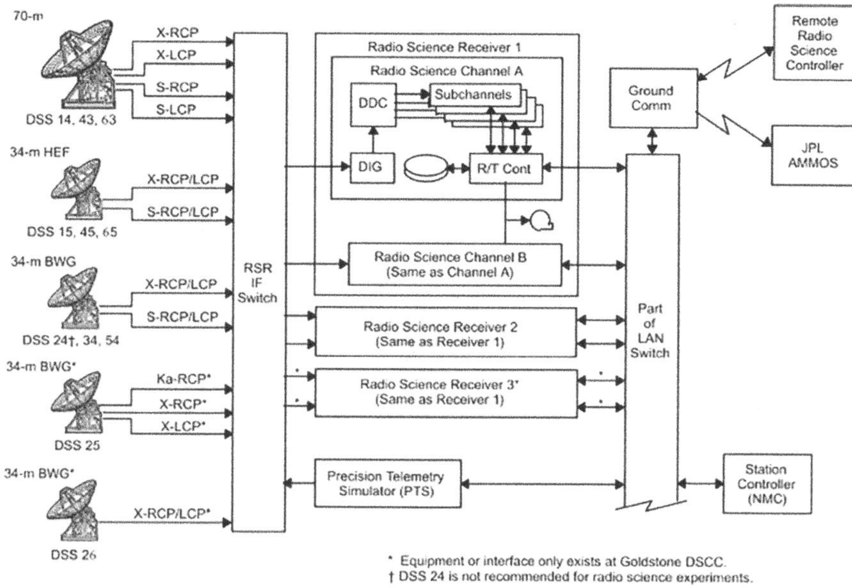

Figure 13.6 Configuration of DSN stations when communicating with Cassini. *Source:* *[8]/TAYLOR, J., SAKAMOTO, L. and WONG, C.J. (2002)/Cassini Orbiter/Huygens Probe Telecommunications./NASA/Public domain.*

antennas. Numerically, the value +9.6 dB was for an uncoded command channel with BER 10^{-5}, and the value +0.31 dB for a telemetry channel with convolutional coding and BER 5×10^{-3}.

13.2.3 New Horizons

Currently, the next most distant interplanetary spacecraft is New Horizons, which was sent to Pluto and its largest moon, Charon, after being gravitationally accelerated by Jupiter [9]. In January 2019, New Horizons flew near another object in the Kuiper Belt of asteroids at the very edge of the Solar System—2014 MU69 (Ultima Thule). The distance of the spacecraft when communicating was 32–40 AU, and the signal returned back in about 8 to 10 hours. The ground segment is provided by the DSN network. Being designed and built by the Applied Physics Laboratory at Johns Hopkins University in Baltimore, the spacecraft is of a somewhat different concept. The block diagram is shown in Figure 13.7. The medium-gain antenna (MGA), used for commanding and regenerative ranging primarily, is a special feature being applied for the first time. The minimum transmission speed is 600 bps. Communication runs in the X band at an uplink frequency of 7 182 MHz and downlink frequency of 8 438 MHz.

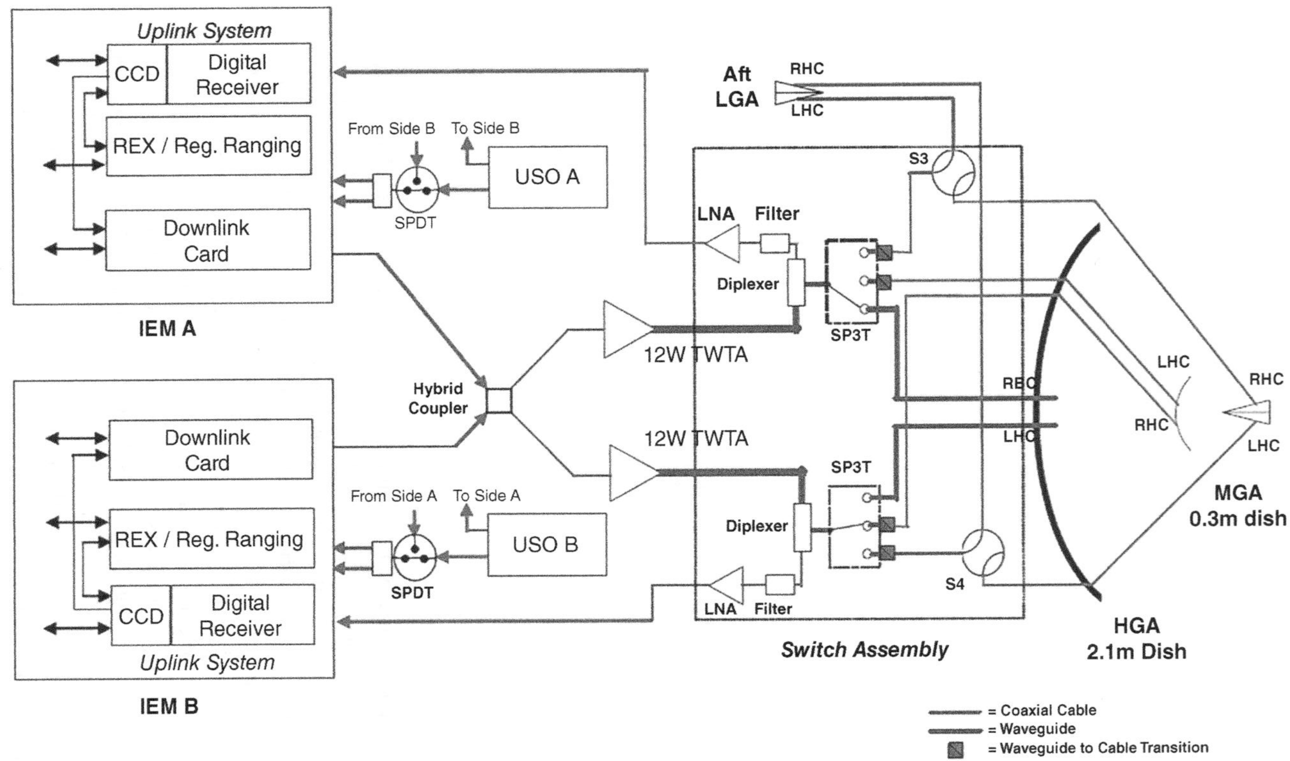

Figure 13.7 Block diagram of New Horizons spacecraft.

13.2.4 Mars Odyssey and Reconnaissance Orbiter

Both these spacecrafts are Mars orbiters, and they had predecessors. The spacecrafts carry a number of instruments for remote sensing of the planet, and serve as radio relays for the well-known research rovers Spirit and Opportunity, and later Curiosity and Perseverance. The Mars Reconnaissance Orbiter [10] can transmit data to the Earth at up to 6 Mbps. The Orbiter has a three-meter reflector. The communication can run not only in the X band but in the Ka band (32 GHz) as well (successfully demonstrated during the flight to Mars). The transmission can run either in real time or from memory (store and forward). The Mars rovers can communicate in the UHF band (400 MHz) with orbiters. The direct communication with the Earth runs at a slow rate in the X band (for command purposes, primarily). Compared to previous spacecrafts, the communication covers *short* distances of about 250 million km with the signal returning in 28 minutes to the Earth.

13.2.5 Spacecrafts at Lagrange Points

There are five Lagrange libration points in the Sun-Earth system. Some of those points are used to locate spacecrafts for Sun research. Most notably, the L1 point lies about 1.5 million km from the Earth on the Earth-Sun line. On the opposite side of the Earth, L2 is at the same distance, allowing a convenient location of a spacecraft for deep space research. The latest James Webb Telescope is placed in an orbit around L2 [11]. L1 is home to the very successful SOHO spacecraft [12], which was launched in 1995 and is currently in operation. Radio communication with spacecrafts in L1 is rather problematic since ground-based highly directional antennas are oriented to the Sun, which is very noisy at radio frequencies. This interference makes communication practically impossible. SOHO (and similar spacecrafts) is therefore moving on an elliptical orbit in a plane perpendicular to the Sun-Earth line with L1 being one of the foci of this orbit. The orbital period is six months, allowing continuous communication with the spacecraft.

In this chapter, we briefly described the DSN radio systems for the communication with very distant and very interesting objects in space. Due to the limited length of the text, a number of interesting projects had to be omitted. Let us name at least the Galileo project to Jupiter, the first Mariner spacecrafts, Rosetta-Philae, and the STEREO project. Those projects are similar from a radio communication perspective since the same DSS technologies and devices are used (which are constantly being improved). On the other hand, the most interesting moments are related to unforeseen events such as the Cassini-Huygens case described above or the STEREO-B spacecraft case [13]. The spacecraft was not able to wake up from hibernation before the conjunction with the Sun. Due to the incorrect orientation of solar cells, the battery was not charging. Gradual (blind) disconnection of spacecraft systems reduced the power consumption. After almost two years, the battery recovered, the spacecraft began to transmit, and control was reestablished. More such situations are known in space communication. Some cases highlight both human intelligence and incredible coincidence.

References

1 Wikipedia. (2023). *NASA Deep Space Network*. https://en.wikipedia.org/wiki/NASA_Deep_Space_Network (accessed 7 August 2024).

2 NASA. (2023). *Deep Space Network Now*. https://eyes.nasa.gov/dsn/dsn.html (accessed 7 August 2024).

3 Wikipedia. (2011). *Beam Waveguide Antenna*. https://en.wikipedia.org/wiki/Beam_waveguide_antenna (accessed 7 August 2024).

4 Otoshi, T.Y. (2008). *Noise Temperature Theory and Applications for Deep Space Communications Antenna Systems*. Artech House. ISBN 978-1596933774.

5 Jet Propulsion Laboratory. *Deep Space Network*. https://deepspace.jpl.nasa.gov/dsndocs/810-005 (accessed 7 August 2024).

6 Ludwig, R. and Taylor, J. (2002). *Voyager Telecommunications*. https://descanso.jpl.nasa.gov/DPSummary/Descanso4--Voyager_new.pdf (accessed 7 August 2024).

7 ESA Communication Department. (2019). *ESA and NASA Agree a New Mission Scenario for Cassini-Huygens*. http://sci.esa.int/cassini-huygens/27650-esa-and-nasa-agree-a-new-mission-scenario-for-cassini-huygens (accessed 7 August 2024).

8 Taylor, J., Sakamoto, L. and Wong, C.J. (2002). *Cassini Orbiter/Huygens Probe Telecommunications*. https://descanso.jpl.nasa.gov/DPSummary/Descanso3--Cassini2.pdf (accessed 7 August 2024).

9 Deboy, C.C., Haskins, C.B., Brown, T.A., et al. (2004). The RF telecommunications system for the New Horizons mission to Pluto. *2004 IEEE Aerospace Conference Proceedings*, Big Sky, MT, USA, 1463–1478. https://doi.org/10.1109/AERO.2004.1367922

10 NASA. (2023). *Mars Reconnaissance Orbiter*. https://mars.nasa.gov/mro/mission/overview (accessed 7 August 2024).

11 NASA. (2021). *James Webb Space Telescope*. https://www.jwst.nasa.gov (accessed 7 August 2024).

12 NASA. (2023). *SOHO*. https://www.nasa.gov/mission_pages/soho/index.html (accessed 7 August 2024).

13 NASA. (2016). *NASA Establishes Contact with STEREO Mission*. https://www.nasa.gov/feature/goddard 2016/nasa-establishes-contact-with-stereo-mission (accessed 7 August 2024).

14

Conclusion

In this book, I have presented interesting radio engineering problems that I had solved when processing very weak signals. As humorously noted by the reviewer, today's radio engineering has condensed into the arrangement *an antenna and an A/D converter behind*. However, this is not the case for weak and extremely weak signals. Although the final processing of signals is digital, the sensitivity of the receiving system is determined not only by the antenna but by other elements of the system as well. Those elements are discussed in the book. In each chapter, I tried to present a brief theory with practically applicable relations and specific solutions of individual parts of the communication systems. Not all of them are listed, but perhaps the most interesting ones have been selected.

When thinking about it, we, along with our peers, were actually extremely lucky for the stage of electronics development we experienced. I started to work with electronics in the era of *tubes* and was involved in the fascinating transition to solid state. Initially, I used germanium low-frequency transistors. In a short time, the first high-frequency transistors appeared and then mesa transistors and silicon high-frequency transistors, usable at hundreds of MHz. Microwave technology with three-dimensional solid waveguides was primarily based on point-contact diodes. Initially, nobody believed that semiconductors would dominate the microwave technology, let alone the high-power high-frequency technology. And what is today's situation? GaAs-based high-electron mobility field-effect transistors (HEMTs) allow the construction of amplifiers, which compete with quantum amplifiers (especially when cooled). High-quality metalized substrates enable the development of planar structures, which are prevalent in the area of microwaves. LDMOS transistors can generate powers of the order of 1 kW at frequencies of several GHz. GaN-based semiconductors allow us to implement amplifiers with outputs of tens of watts at frequencies of the order of 10 GHz. Thanks to technologies of integrated circuits, a high-quality microwave frequency synthesizer can be implemented on a single chip (and therefore, in a single package).

Reception of Weak Radio Signals from Space, First Edition. Miroslav Kasal.

I was also greatly influenced by the amazing development of digital electronics: from the first TTL gates, through CMOS technology, the first four- and eight-bit computers to today's 64-bit PCs with clock signals in the GHz range. In the field of signal processing, I witnessed the development of signal processors and programmable gate arrays, including the development of sophisticated sampling and encoding and decoding techniques. The development of computers was followed by intensive development of application programs (ANSYS, CST Studio, PSpice, Matlab, and others) that allow us to perform previously impossible computations and simulations.

The community of people involved in radio astronomy, in communication by reflection from the Moon, and in receiving signals from distant satellites and spacecrafts includes both professionals or previous professionals just in the field or in some neighboring disciplines, and amateurs. Their common characteristic is curiosity, the ambition to *hit the nail on the head* and to achieve top parameters of their systems. In any case, experimental work in this field is a great opportunity for self-education, acquiring skills, and new, previously unknown knowledge. I would be pleased if this book could stimulate readers' interest in the issue of weak signals I am still fascinated by.

Index

Printed and bound by CPI Group (UK) Ltd, Croydon, CR0 4YY
17/09/2026
14964333-0001